Tapping into Water

Low Tech Well Drilling Techniques and Tools

This publication describes the general procedures for fabricating tools, drilling, and building wells. The process of building anything depends largely upon the skill of the builder, the materials used, the machinery employed, and the conditions under which the work is performed. The author cannot and will not assume any liability for damage to persons or property or other consequences of any procedures or designs referred to herein or of any omissions relating to techniques, design, materials or procedures. Construction of wells is regulated by your state or local municipality. Fee's, permits, and permissions are assumed to be a prerequisite of drilling. Unfinished bore holes must always be covered in a child-proof manner, and abandoned bore holes permanently filled. Any water derived from a naturally occurring source for the purpose of human consumption must be tested initially and regularly thereafter for impurities. The author makes no claim of ownership or novel invention regarding the construction methods shown or the designs presented in this publication.

Tapping Into Water Low Tech Well Drilling Techniques and Plans

Copyright © 2010 Paul Sawyers

www.paulsawyers.com

Contents

Preface... 5

Chapters

1. Percussion Drilling to Create a Well Hole........................ 7

2. Sludging (Reverse Jetting) to Create a Well Hole.............. 61

3. Hand Auger to Create a Well Hole................................ 71

4. Drive Points for Creating Ready Made Wells 81

5. Well Installation; Casing & Finishing Bored Well Holes 91

6. Water Flow Development of New Wells 119

7. Hydrology Basics & Locating Ground Water................... 123

Preface

The aim of this book is to present a handful of time tested methods for well hole boring coupled with modern methods for well finishing. These primitive well drilling techniques and new variations on them can be manufactured and used by regular people without a great investment of income or specialized tooling.

In general, most ground water lies within the first 200 feet and the majority of wells are only 50-100 feet deep. Of course ground water occurs at much greater depths, but not in the usual places people choose to drill for water and inhabit.

It's expensive to have a well drilled by a professional, and say for example the bill is $10k, this price does not usually guarantee water if the driller creates a dry hole.

The fiscal pressure for success is greatly reduced when well sinking is undertaken by the part time handyman or weekend tinkerer. Time and a small amount of money are the only thing at risk in this endeavor. Upon successful completion of your own well, you can often re-coup spent funds by selling the rig, or you can keep it for future use and loan the fabricated well drilling tools and equipment to neighbors.

A simple way to get a jump on the feel for boring a proposed well can be achieved by visiting your local hardware or home improvement store and purchasing a clamshell style post hole digger. Take this tool to the spot you plan to drill and dig a small diameter (usually 6-10 inch) post hole. Go as deep as possible. After about 3 feet your progress will halt as you find the handles can no longer open wide enough for the digger to function properly. You may also encounter some rocks or clay.

This post hole digging exercise is cheap and gives you an idea of what type of materials exist in your desired drilling area. It also demonstrates the basic premise of well drilling; digging tool goes in - the material is lifted up and out to create a small diameter hole. The limitations of digging tools are also illustrated as the clamshell post hole digging apparatus almost always fails at the 2-3 foot point.

Above: clam shell style post hole digger reaches it's depth limit at 2-3 feet.

The act of loosening and turning up the ground formation in a controlled small diameter bore hole, then removing this loose ground formation material from the bore hole, is the primary act of well drilling.

Most of the tools shown in this book rely on vertical movement and vertical operating apparatuses - thus avoiding the horizontal opening and closing operation of the clam shell blades as described.

Take a length of steel conduit or copper plumbing pipe outside and plunge it into the ground. What remains upon withdrawal is a plug of dirt inside the tube. This is the same vertical cutting motion many old style drilling systems rely on and this is the best premise on which to build low tech, low cost rigs.

The goal of this book is to revive a few of these choice technologies in a modern format that can easily be reproduced with common materials and basic skills.

Percussion Drilling to Create a Well Hole

Overview

One of the oldest techniques used in deep well construction is percussion drilling, or cable-tool drilling. While in essence the technique has always been a percussive movement (where the cutting tool impacts at the bottom of the bore hole in a chopping action), it took the name 'cable-tool' much later when giant bits and motorized cables (invented during industrial revolution) became the standard mode of operation.

Traditional percussion drilling on a smaller scale usually involves one or more people operating a small rope and pulley type rig. There is no gasoline or electricity required for this kind of operation, just human power. The trade off is operator fatigue and slow progression of work. Drilling a 100-200 feet deep bore hole can take several weeks using the percussion method, but shallow wells of 25 +/- feet can often be bored with just a few days of work.

While it's common to run a small percussion drilling rig with a crew of two or more people, it is feasible for a single worker to operate it. Low tech percussion drilling can penetrate a wide variety of ground formations, so it is not limited to a certain soil type like many other low tech drilling methods. As you will note, this chapter on percussion takes up more pages than the others, simply for the fact that it is such a universal method deserving of more attention (photos, diagrams etc).

The basic operation of the percussion technique is achieved by a heavy steel cutting bit, usually 3-6 feet long, attached to a rope and pulley. The rope is pulled and manipulated above ground resulting in the up and down movement of the bit at the bottom of the bore hole. The bit repeatedly impacts the bottom of the bore hole with a chopping action that cuts into and mixes up the ground. The bottom of the bore hole gradually accumulates a small level of drilling mud during this cutting action, created by the operator pouring a bucket of water into the hole every so often. This viscous mud acts to suspend and lift the cut materials for removal by the bailer.

When a few feet of drilling mud accumulates, (indicated by the mud line on the bit) the cutting bit is raised out of the hole and replaced with the bailer. The bailer is a steel tube usually 3-6 feet long that fits into the bore hole. The tip has a valve that allows mud in capturing it for removal.

As the bailer is rapidly lowered into the bore hole and impacts the bottom, the valve opens, captures the mud, then the bailer is raised up and out of the hole for emptying on the surface. The bailer is lowered again to remove any remaining mud. Once the bailer comes up empty, the cutting bit is replaced.

Access to a welding shop or hiring a small welding shop to fabricate the tools shown in this chapter is necessary as all the bits and bailers are made from steel parts. Chisel edges on cutting bits can be re-ground sharp and the tools stored for future use, sold, or loaned out after your own well is finished. The price paid for fabrication is money well spent when you consider that these tools may produce a dozen wells or more before being broken or retired.

The low tech percussion rig is comprised of two main aspects:
1) above hole tooling (rope, tripod, pulley, human power, etc)
2) below hole tooling (steel cutting bits, steel tubular bailers with valves)
Fabrication and operation of these aspects will be explained further in this chapter. Photos and diagrams will be an essential part of this explanation.

Above-Hole Tooling

Low tech percussion drilling relies on a tripod setup directly over the bore hole to suspend and manipulate the various tools suspended by a rope.

The main above hole components of a low tech percussion rig are: the tripod, pulley, and rope. Rope should be anchored to a deep stake, tree, vehicle, or a similarly secure anchoring point.

To supplement this basic array of tools you can add a fulcrum, lever, and a winch. The fulcrum and lever can be used as an alternative rope attachment point providing a rapid up and down-short distance cutting action to the bit. The hand winch can be mounted to a tripod leg for lifting heavy bailers filled with material, or pulling out bits stuck in mud. The rope is pulled to lift the cutting bit a short distance (1-6 feet) using a handle or lever wrapped around the rope, or by pushing and pulling the rope up and down by hand (more on operating procedures later in this chapter).

Essential Above Hole Components of Low Tech Percussion Drilling

º **9-12 foot tall tripod made from metal tubing or 4x4 lumber.**

º **6-18 inch pulley or bicycle tire rim rigidly mounted to the tripod center.**

º **Rope anchor point 50-100 feet from bore hole (tree, stake, parked vehicle).**

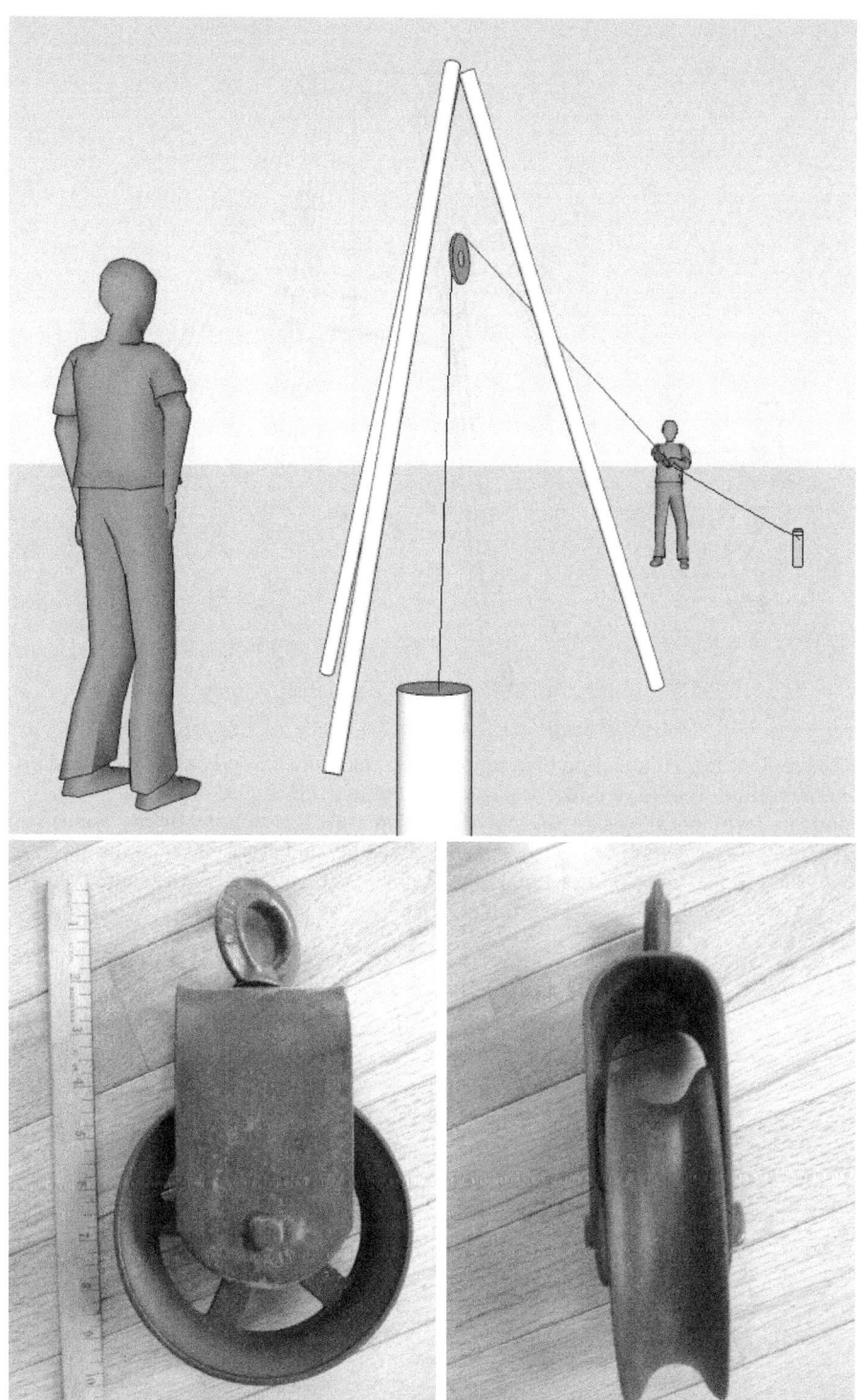

Top: tripod, anchored rope, human powered drilling.
Bottom: 6 inch diameter farm pulley.

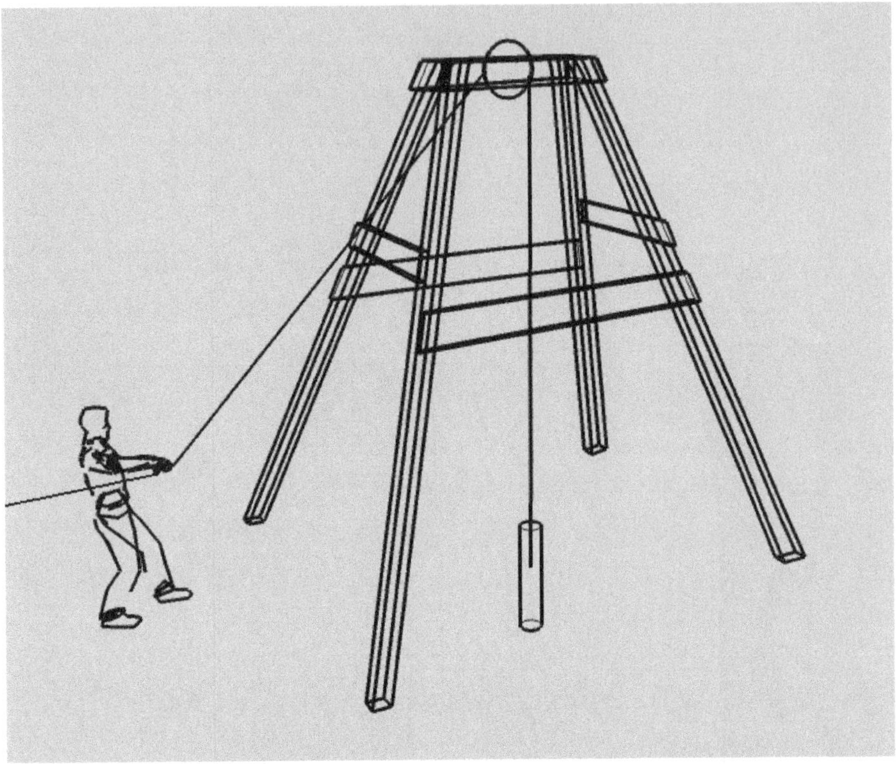

Above: four legged 'quad-pod' variation of the traditional tripod made with standard lumber. Built from (4) ten foot 4x4's and (6) ten foot 2x8's. Use of carriage bolts and large washers allows disassembly for transporting and storage. Below: welded plate steel body can be fabricated for quick assembly and disassembly of the 4x4 legs used in the "quad pod" setup. Pulley can be center mounted with a large bolt.

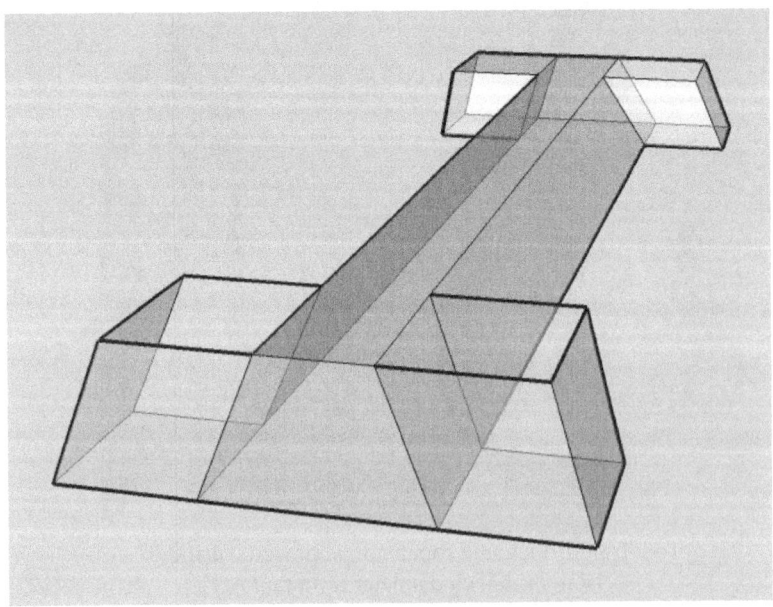

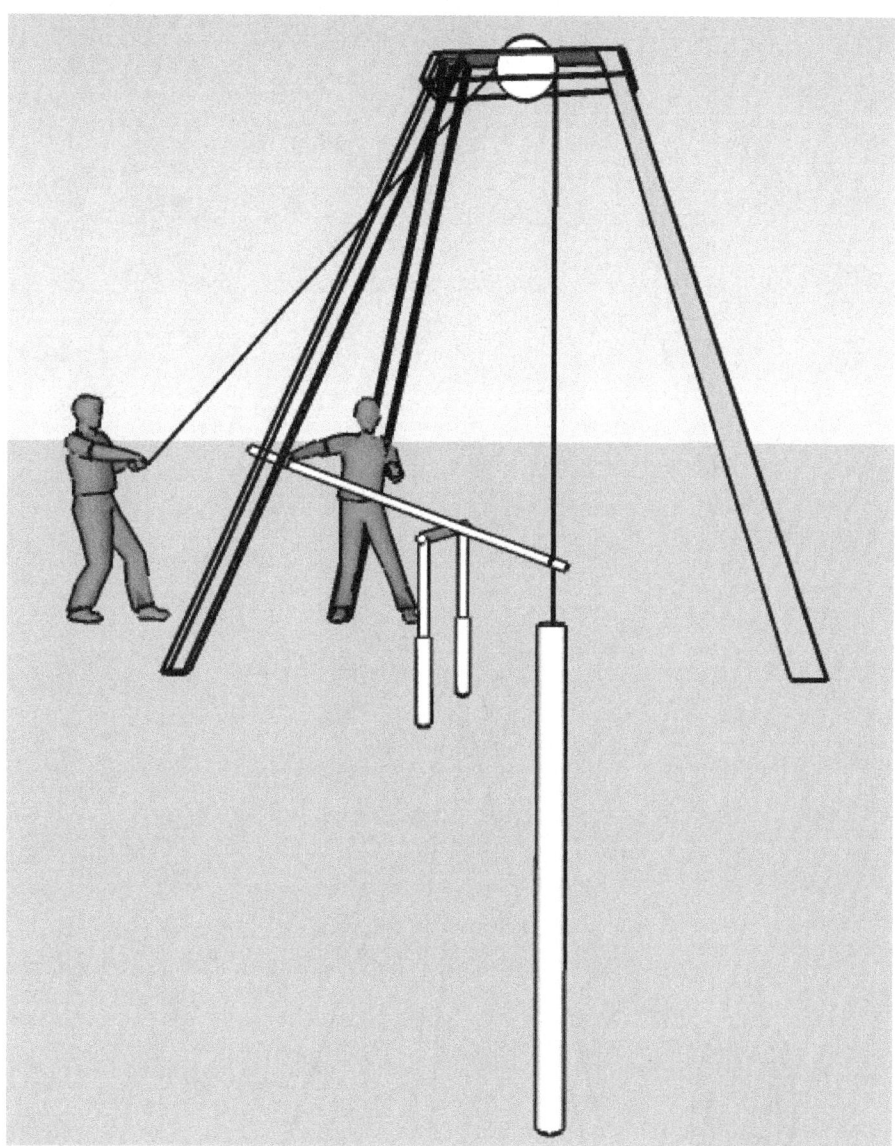

Above: the fulcrum and lever is a simple addition to the percussion rig consisting of two upright posts mounted in the ground with a horizontal beam bridged across both. This set-up is useful for generating repeated short distance vertical percussion movement. The leverage bar utilizes downward applied pressure which is beneficial as an alternative to the overhead rope tool (that relies on pulling force). Changing body positions is often helpful when operator fatigue sets in. Since the fulcrum and lever is usually only used temporarily during operation, a few simple loops of the existing rope (around the lever) is often enough to provide a rapid and easily removed attachment to the in-hole bit. When finished, the lever and horizontal beam can be removed and set aside until needed again.

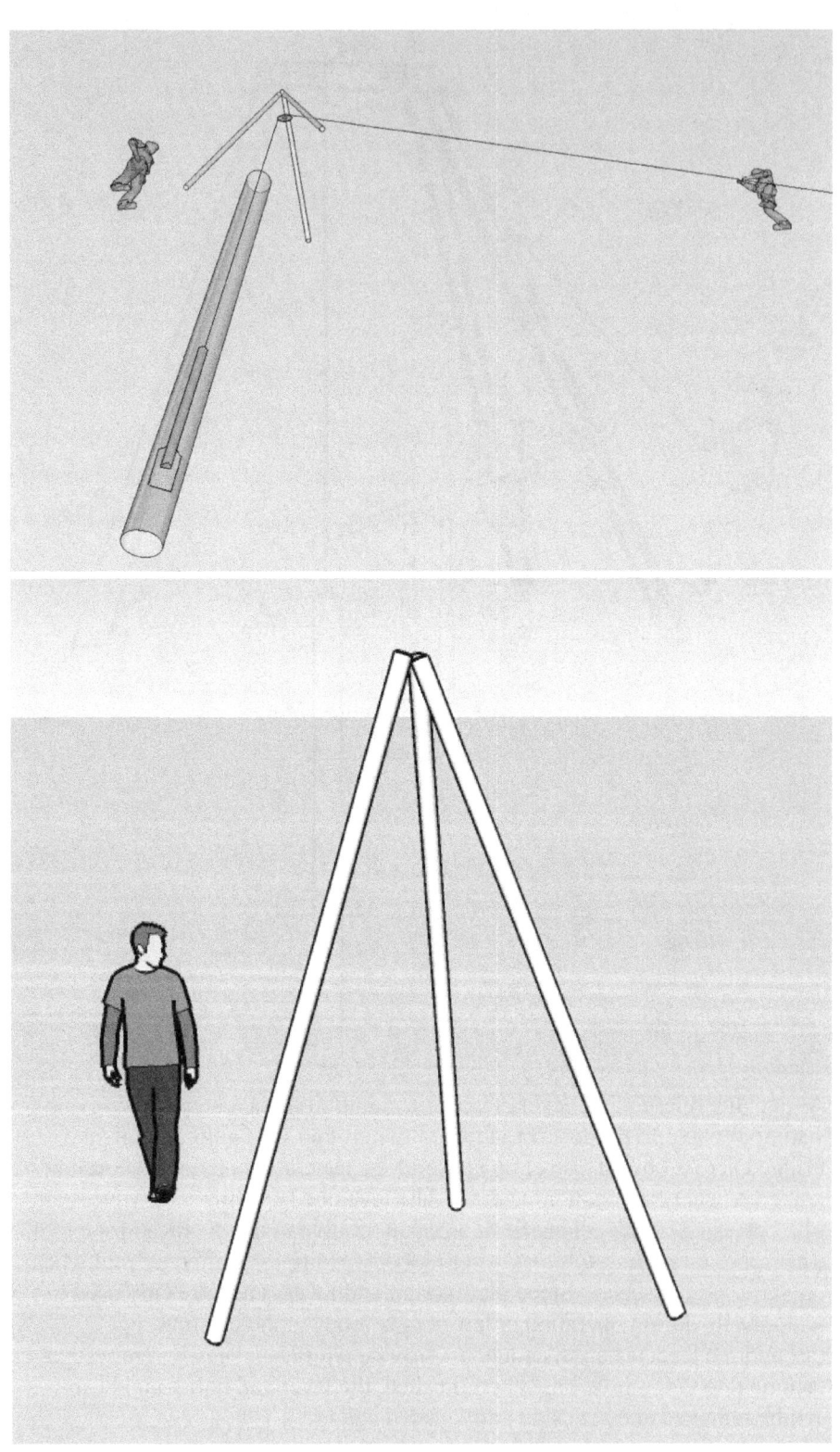

Below-Hole Tooling

The two main below-hole components of a low tech percussion drilling rig are the cutting bit and bailer. While one multi purpose star shaped cutting bit is often enough, usually a few designs are used. The bailer is a pipe with a valve or flap at the end lowered into debris filled bore holes to remove cuttings. The bailer and cutting bit of a percussion drilling set should be roughly the same diameter, but the set itself is fabricated to fit the size of the hole desired, with bore holes of 4-8 inches in diameter being most common.

Before you start to fabricate percussion bits for a small rig you should calculate the weight of the steel used. One piece of round steel stock 2 to 4 inches in diameter and 3 to 6 feet long with a simple chisel ground into the tip and a steel loop welded to the other end will provide ample heft. This is a percussion bit, a very simple one, but a functional bit none the less. The combination of weight and an aggressive plate steel cutting blade design will penetrate most ground formations including rock, leaving broken fragments and mud for the bailer to remove. Keep in mind that there is a limit to how heavy a bit you can realistically operate above ground by hand. 30-50 lbs is a good place to start for one person. More rapid progress is made when you recruit helpers to operate a 80 lb or heavier bit. Large bailers full of cuttings can also become too heavy to lift by hand, one reason why winches are often used with small percussion rigs.

While round bar stock over 3 inches in diameter can be had, it is not as easily found (kept in stock) as 3 inch and under. With this in mind you can effectively utilize 2 inch, 2.5 inch, and 3 inch steel round bar stock for boring 4-8 inch diameter well holes. Various sized cutting blades are welded to the working end of the bar stock to achieve the desired cutting diameter for the bore hole needed. You can also weld a cluster of several smaller diameter pieces of round stock together to create a larger heavier tool with a quad array of cutting blades.

Since the body of the percussion tool merely provides weight, just about any large steel object can work provided it's close to round in shape and a well designed cutting blade is welded to the percussion end (3 foot long piece of railroad track for example, with a large four head cutting blade fabricated from 1/2 inch plate steel).

It is important that you always use the longest heaviest piece of steel that you can comfortably operate when creating the body of a percussion bit. Bit bodies can always be cut shorter later if you discover it is too heavy to lift and operate. Weighted extensions can also be welded on if you find the bit is too light to effectively penetrate ground formations.

Cutting blades should be made of plate steel, a minimum of 1/4 inch thick, but preferably 1/2 - 3/4 inch thick. Cutting blade surfaces are usually ground to a chisel point to maximize impact efficiency. Repeated use of percussion bits in rocky terrain will eventually dull cutting blades, but you can re-sharpen the edges when needed using an angle grinder.

1018 Round Bar Weight

Diameter Stock	Weight Per Foot
2 inch	10.63 lbs
2.5 inch	16.66 lbs
3 inch	23.99 lbs

A36 Round Bar Weight

Diameter Stock	Weight Per Foot
2 inch	10.70 lbs
2.5 inch	16.72 lbs
3 inch	24.08 lbs

Salvaged steel can sometimes be used for fabricating low tech drilling tools but keep in mind the fact that it is usually sold to steel recyclers and not the general public, which is why you may not often find steel stock at a salvage yard. Mechanical shafts, hydraulic pistons, and so forth can be found and adapted for use as percussion bit bodies. There are many variables in what you may find at a scrap yard, so many that listing possible usable objects here is likely an exercise in futility. You might consider first looking around at your local salvage yard to see what you can find for cheap, and then if needed, purchase the new steel stock. Note: many salvage yards have liability policies in place the prohibit customers from picking out material themselves.

Salvaged round bar material can be used, and is often (from a fiscal mindset) preferred. Many times finding what you need in a salvage yard at the same point you need it are two events that fail to coincide. With this fact in mind purchase of raw steel stock should be considered. New steel stock may not cost much more than what you are charged at a salvage yard.

Any monies spent on raw materials can easily be recouped with the first successful well hole bored and cased, then your percussion bits can easily be sold, rented, or lent after you are finished using them.

If you go the route of purchasing new steel round bar stock for bit fabrication, there are two main types of steel you will probably utilize. These are 1018 cold rolled steel, and A36 hot rolled steel. Both are frequently used types of mild steel but offer slightly different characteristics.

1018 is the most commonly available of the cold-rolled steels. This is a good steel for mechanical applications such as bending and welding. It has a smooth surface created from the cold rolling process that produces minimum abrasion for rope attachment loops etc.

A36 steel is a hot-rolled steel. It's the most common type of structural steel used in the United States. The hot roll process creates a surface that is somewhat rough.

Generally, the stock shapes used for fabrication are round, and plate steel.

Raw steel stock can be purchased in most urban areas from local steel vendors where you can choose from round bar and plate steel in all shapes and sizes. Look under 'Steel Distributors & Warehouses' in the phone book to find these local outlets and inquire further about steel purchases.

Left: 1018 steel lends itself well to the fabrication of 1/2 inch diameter bit hangers that you will see in the coming pages. These are the steel loops welded to the back end of percussion bits and bailers for connecting the rope.

Common Percussion Cutting Bit Dimensions

Percussion cutting bit blades should be designed for the ground conditions you will be drilling. Flat chisel style blades work well for breaking rocks, and four blade designs are good for regular soils. These are most common, but bit design is flexible and exact dimensions used will vary.

Below left: chisel bit 4.5 inch wide blade.

Above right: star bit 7 inch wide blade.

Tool Weight	a	b	c
35 lbs	4-8 inches wide	3 foot length	2 inch diameter
68 lbs	6-8 inches wide	4 foot length	2.5 inch diameter
96 lbs	6-10 inches wide	4 foot length	3 inch diameter

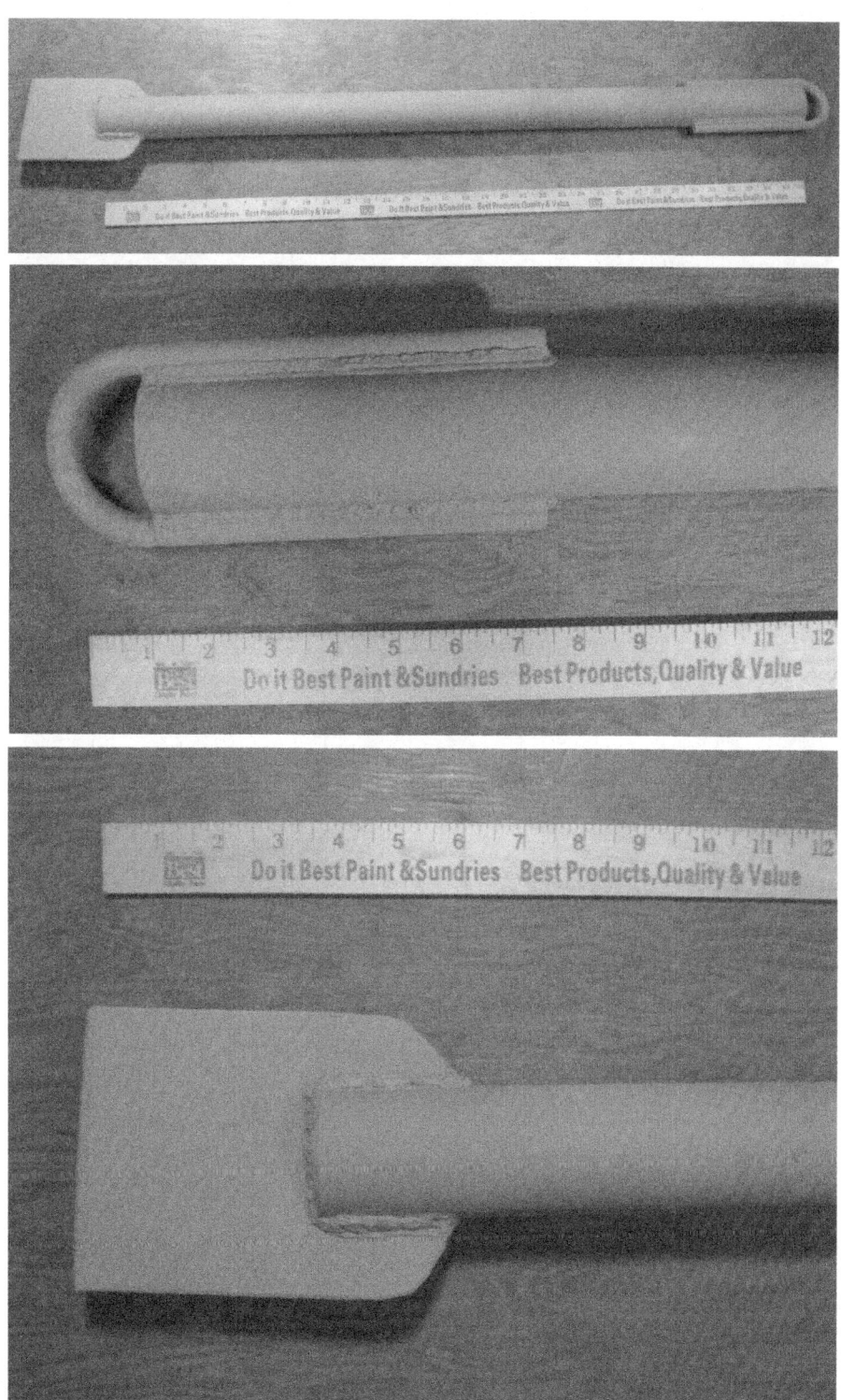

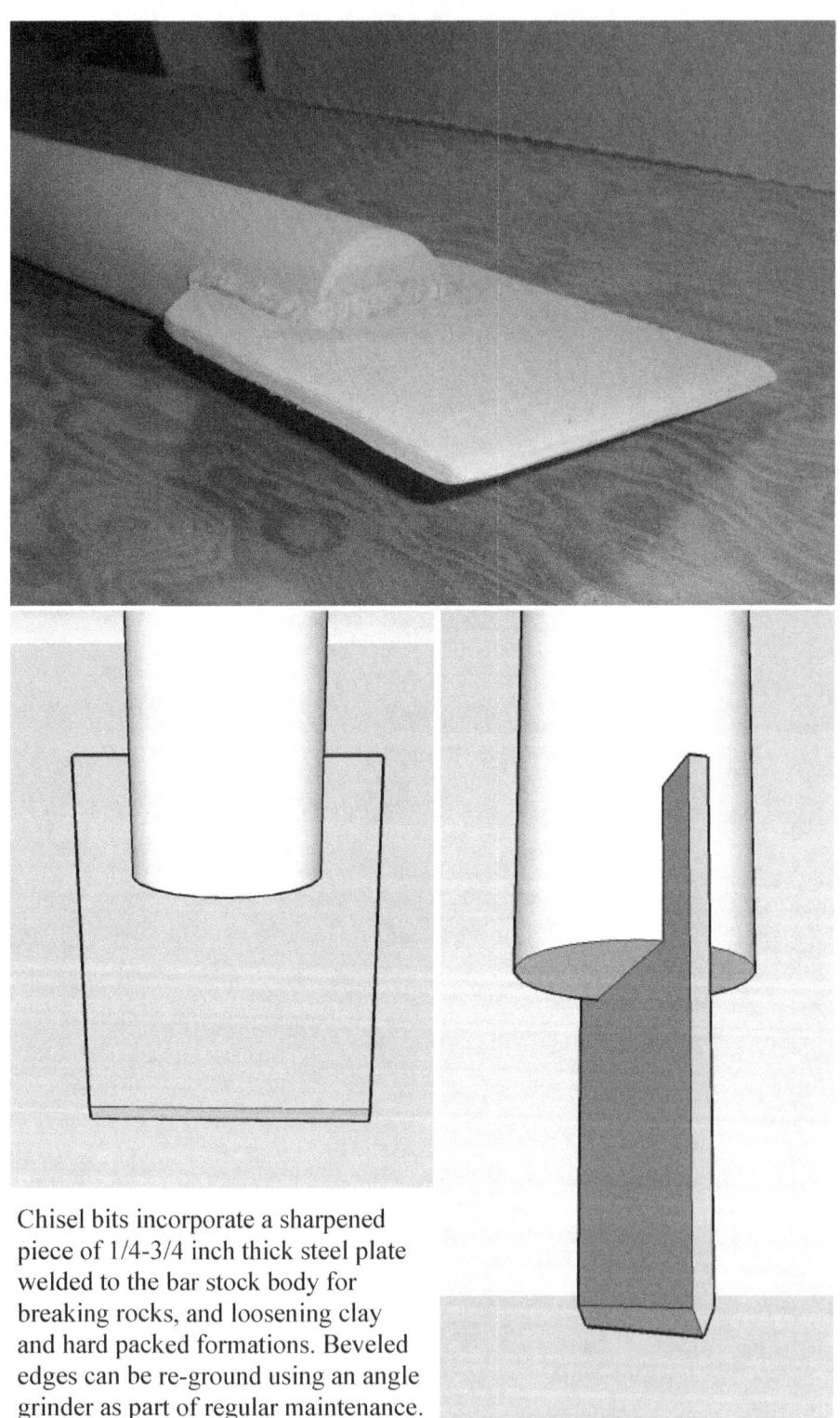

Chisel bits incorporate a sharpened piece of 1/4-3/4 inch thick steel plate welded to the bar stock body for breaking rocks, and loosening clay and hard packed formations. Beveled edges can be re-ground using an angle grinder as part of regular maintenance.

Variations of Star Blade Percussion Drilling Bits

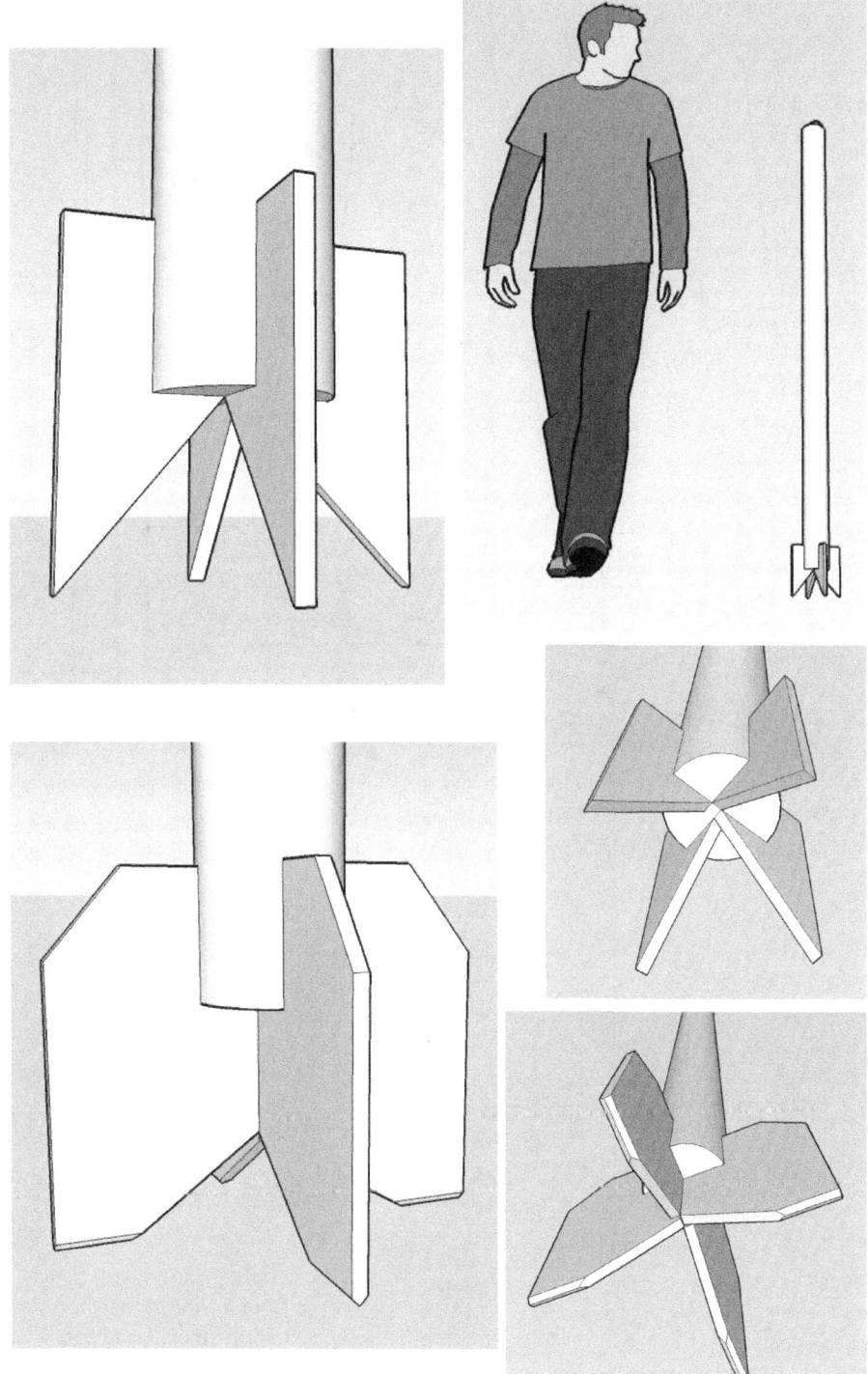

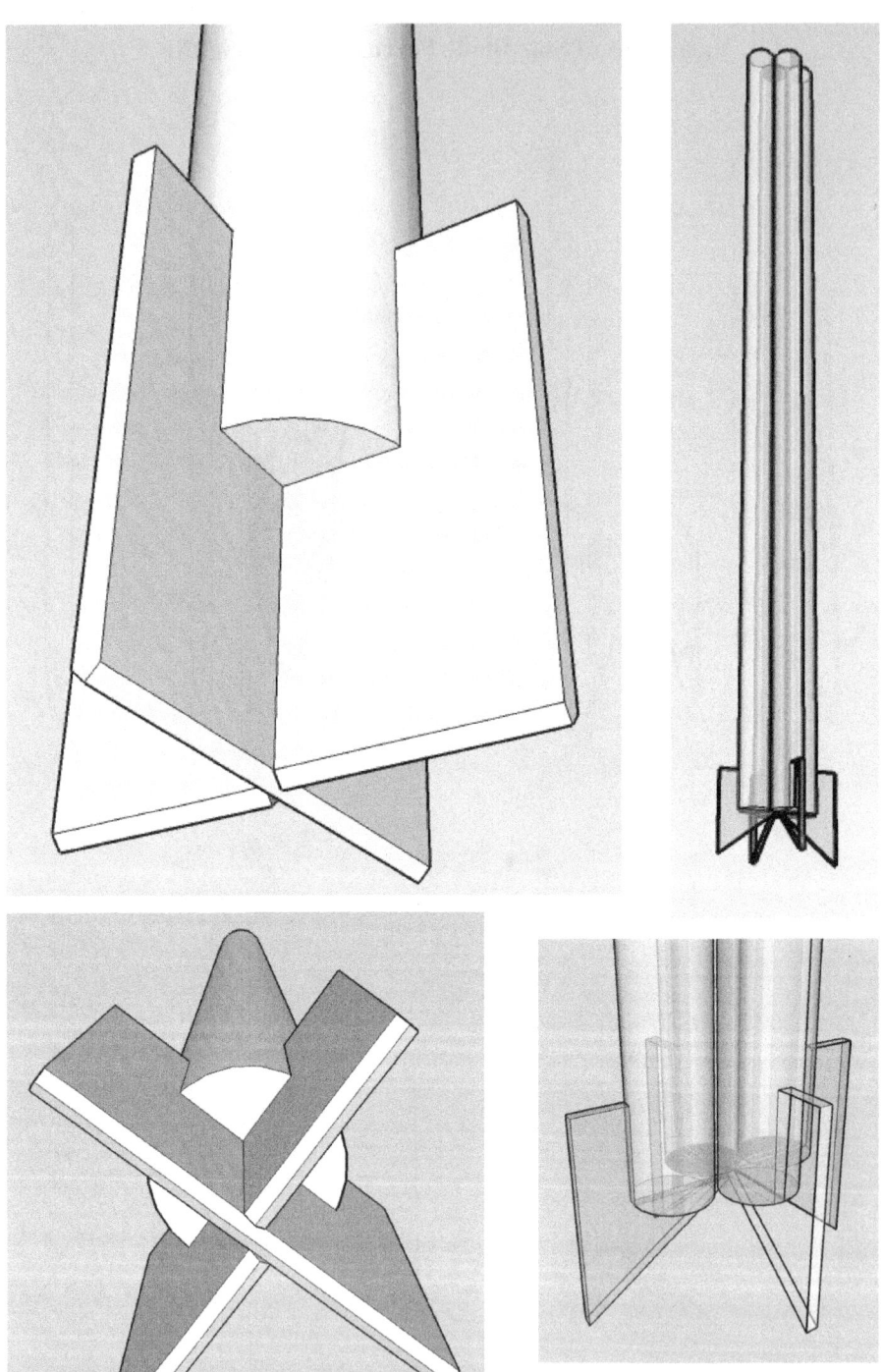

Above right : star bit with body made from a welded together cluster of one inch round bar.

Above left: chopping action flat faced star bit.

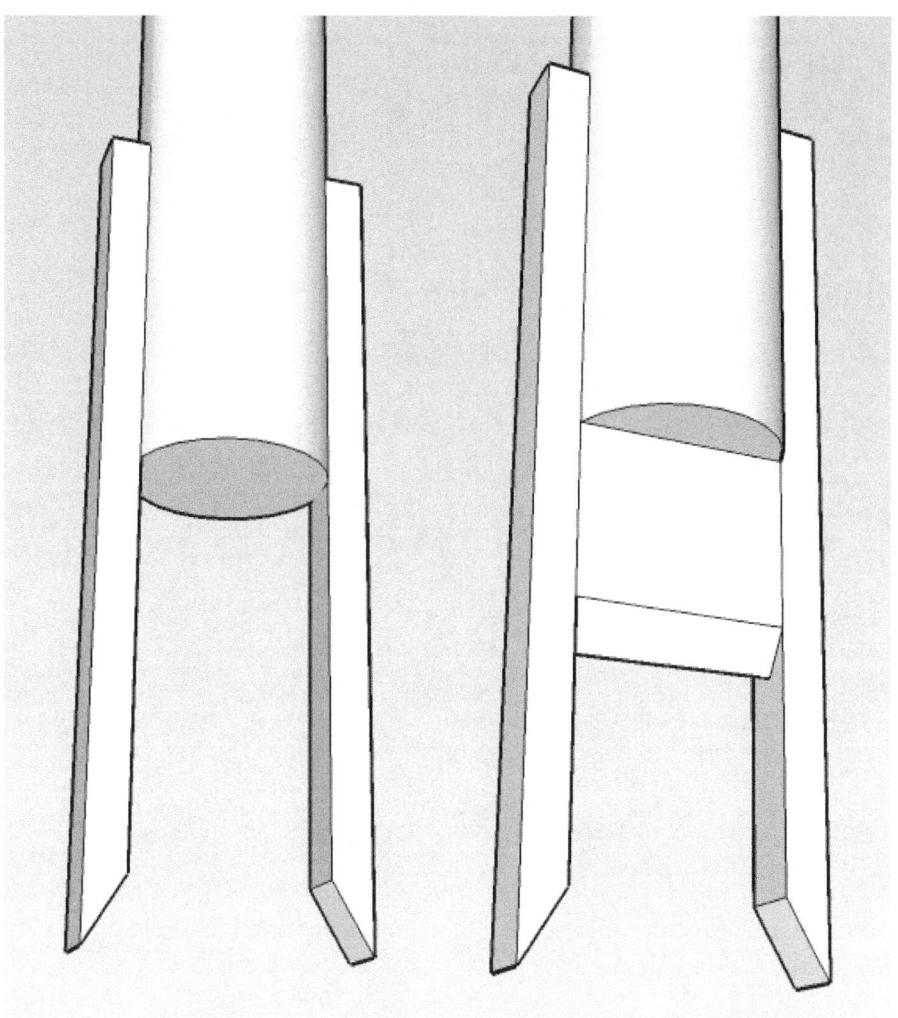

Designing Custom Percussion Bit Variations for Special Situations

There are a variety of ways to build percussion bits with mild design changes that result in improved ground condition performance. Round or square steel stock is usually a basic component that forms the framework for expansion with additional welded-on aspects. The prototype fork style designs in these images offer additional bore hole chopping penetration through hard packed formations.

Ideas for bit design can be found by studying common hand tools that utilize a downward vertical chopping action (like an axe). Often times a required bit design will not be known until the drilling has already begun and a certain underground formation has been encountered. These specialized bits can be designed and fabricated on-site or at the closest available welding shop.

Left: star bit with four main 1/4 inch wide plate steel blades accompanied by four secondary 1/4 inch steel inner blades produces a more efficient cutting action and a stronger bit.

Bailers

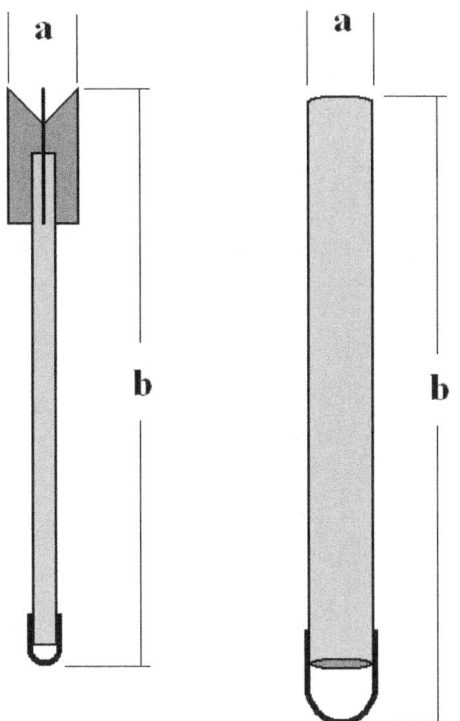

The bailer is the second component of below hole tooling. It carries out the task of removing the debris created by the cutting bits.

Drilling mud, fluid, and small rock fragments can all be successfully removed by the bailer.

There are two main types of bailer: 1) the dart valve bailer, and 2) the flap valve bailer.

Both styles will be illustrated in the coming pages and both feature a similar tubular body as seen in figure -b- to the left.

The above diagram shows the relation in dimension of cutting bits and bailers with figure -a- representing width of tools (and ultimately the bore hole width), and figure -b- representing the length of tools.

The width of the cutting blade used closely equals the resulting width of the bore hole. The width of the bailer tube used should be the same diameter as the cutting bit. For maximum bailing efficiency, the tube should be large enough to make a full spanning sweep of bore hole debris when sent to the bottom. This means it should be just about as wide as the bore hole but not too large that it's free movement is restricted by contact with bore hole walls.

The length of cutting bits and bailers are, for the most part, unrelated. Cutting bits will usually be in the 3-6 foot long range with a predictable pre-determined weight. Bailers on the other hand have two weights; empty when lowered and full when removed. So figure -b- is important for the bailer design itself and usually lies in the 3-6 foot long range.

Tubular Bailer Body

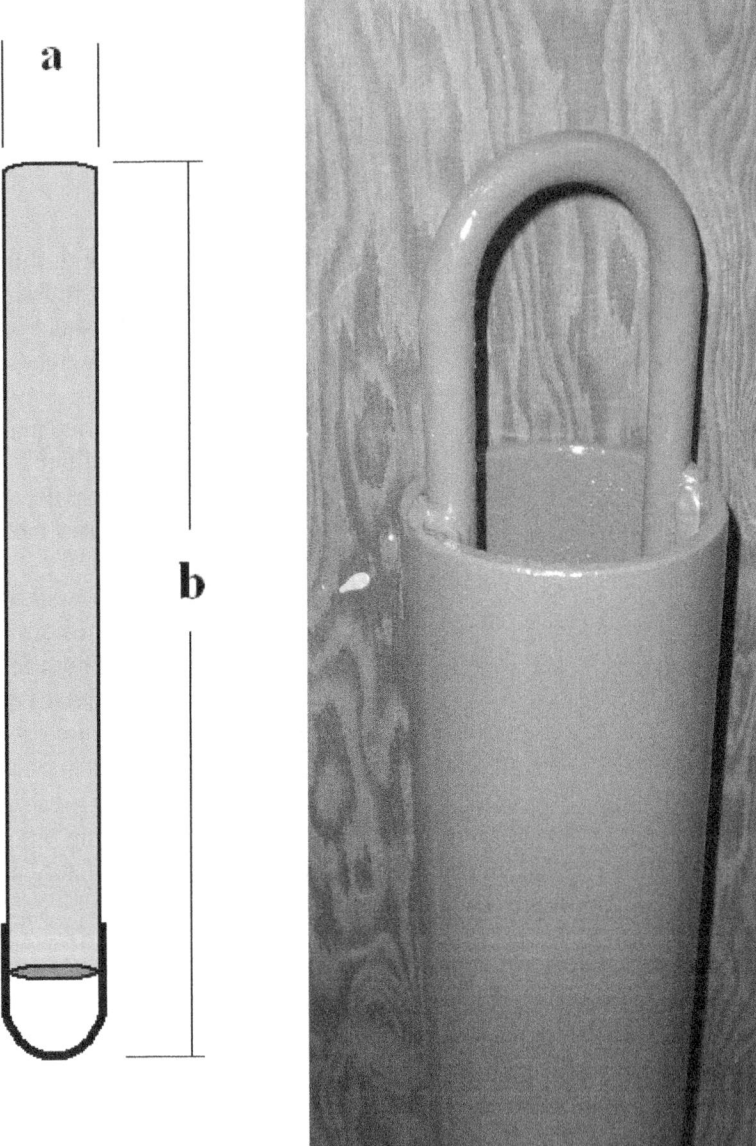

Tubular bailer bodies can range from 3-8 inches in diameter, and are often 3-6 feet long. The best bailer is one you can handle when loaded. If that means you can only use a 3 foot long bailer sized for the bore hole diameter, so be it. There is a limit to how much material laying at the bottom of a bore hole the cutting bit can penetrate, so you may have to haul up much less than a full bailer load just to clear a path for further cutting. Half inch round bar stock is bent and welded to the inside wall of the tubular body creating a rope attachment point for lifting the bailer.

Tubular Steel Stock

Choosing between steel pipe and steel tube when fabricating a bailer body often boils down to what's available (style and size), and price. The choice ultimately falls on the builder to decide what dimension and type of pipe or tube will work best with their bailer design.

Pipe is somewhat more rigid than tube, and is produced in heavier wall thickness. Since the main purpose of pipe is transporting fluids or gases, the most import property is the capacity or the inside diameter (ID).

Pipe is specified by a nominal dimension which bears little resemblance to the actual dimensions of the pipe. It is common to identify pipes in inches by using NPS or "Nominal Pipe Size". The metric equivalent is called DN or "diametre nominel"

Since the outside diameter (OD) of a single nominal pipe size is kept constant the inside diameter of a pipe will depend on the 'schedule' or the thickness of the pipe. The schedule and the actual thickness of a pipe varies with the size of the pipe. Higher schedule number equals thicker wall.

The tolerances are looser for pipes compared with tubes and they are often less expensive to produce, but not always less expensive to buy.

Tube stock is measured by nominal dimensions that are based on the outside diameter. The inside diameter of a tube will depend on the thickness of the tube (Wall). The thickness is often specified as a gauge. The tolerances are higher with tubes compared to pipes and tubes are often more expensive to produce than pipes, but not always more expensive to buy.

Product Type Examples	Outside Diameter (OD)	Wall Thickness	Inside Diameter (ID)	Weight Per Foot
4" Schedule 10 Stainless Pipe 304/304L	4.5"	0.12"	4.26"	5.7 lbs
6" Schedule 10 Stainless Pipe 304/304L	6.625"	0.134"	6.357"	9.14 lbs
3" A513 Type 5 Steel Tube	3"	0.1875"	2.625"	5.64 lbs
3.5" A513 Type 5 Steel Tube	3.5"	0.375"	2.75"	12.50 lbs
4" A513 Type 5 Steel Tube	4"	0.083"	3.834"	3.45 lbs

Dart Bailer

The dart bailers name comes from the dart-like valve at the inlet end of the tubular bailer body that opens to allow drilling mud, fluid, and small rock fragments in on the downward stroke, then closes on the upward withdrawal stroke trapping this material inside the tubular body for removal from the bore hole.

Dart bailers valve bodies are round steel cup-like discs that fit into the end of the bailer tube. Bodies and dart assemblies can be custom fabricated in a small machine shop using a lathe, torch or plasma cutter, and welder. Dart bailer valve bodies can also be fabricated easily by welding a few common pipe fittings together as shown on the following pages. Do-it-yourself dart assemblies can also be made with common materials such as bolts, reducer couplers, and washers.

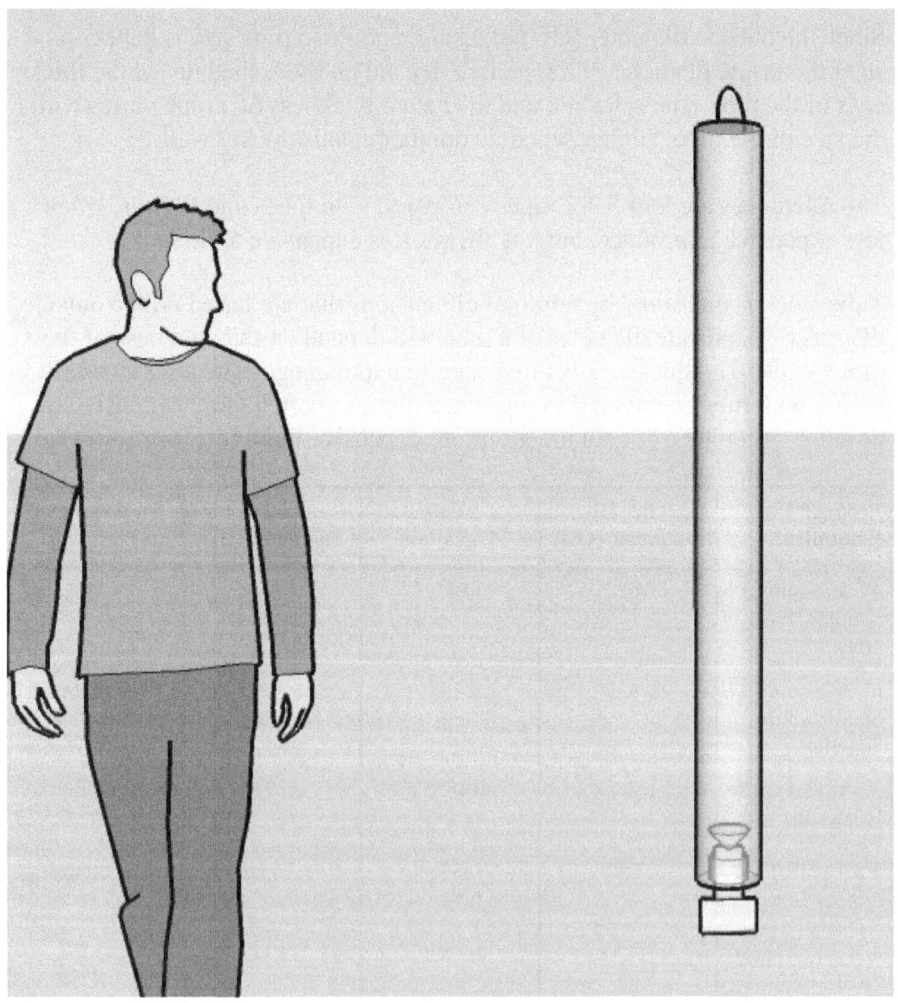

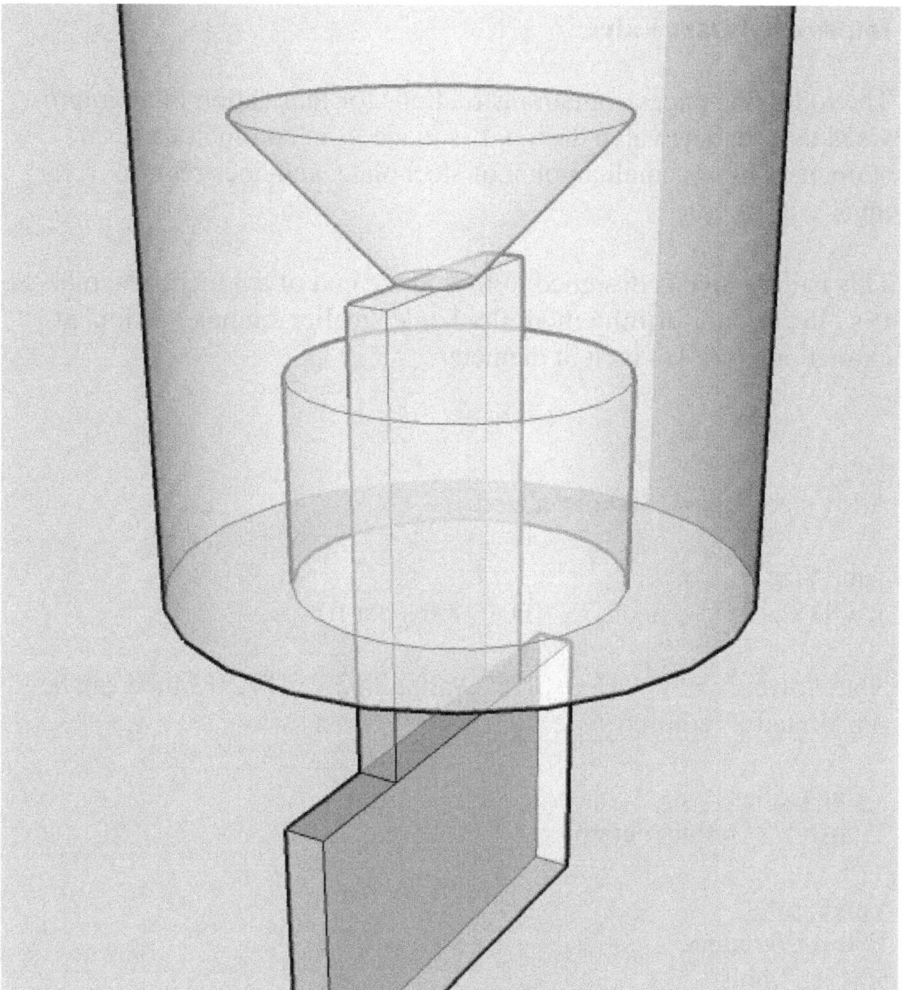

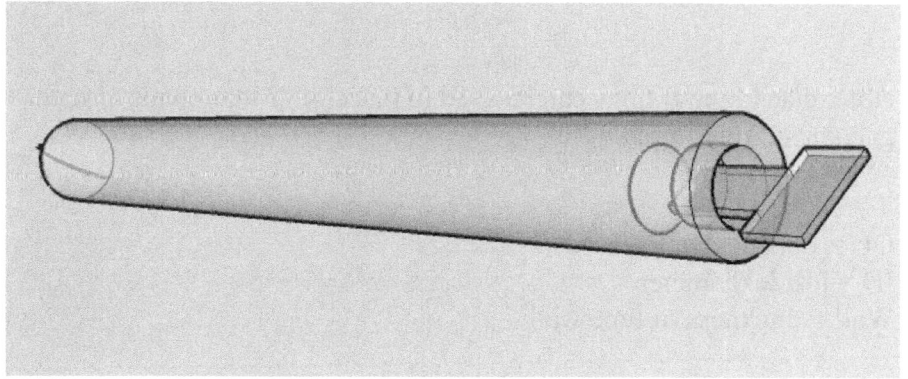

Improvised Dart Valve

The following pages contain instructions for fabrication of an improvised dart (bailer) valve design. It's made from common hardware store parts plus a small amount of steel plate, and does require welding.

This bailer valve is designed to fit into the end of a 3 inch steel tube. It's effective for drilling mud, fluid, and smaller diameter sediment extraction up to 1/4 inch in diameter.

Tube
Alloy steel tube 4130 normalized

Tube Size
3" OD x .125" wall x 2.75" ID (3.8 lbs per ft)

Note: lower cost 3" OD x .1875" wall x 2.625" ID A513 tube can be substituted if required

Valve Collar Body
2" to 1-1/2" reducer coupler

Valve Parts
1" to 1/2" reducer
5/8" x 5" bolt
5/8 large washer
5/8 lock washer (bent flat)
1/4" thick x 4" long x 4" wide 1018 flat bar stock (for dart bailer blade)

Note: black pipe fittings can be used in place of or interchanged with galvanized if required

OD = outside diameter
ID = inside diameter
Wall = thickness of tube wall

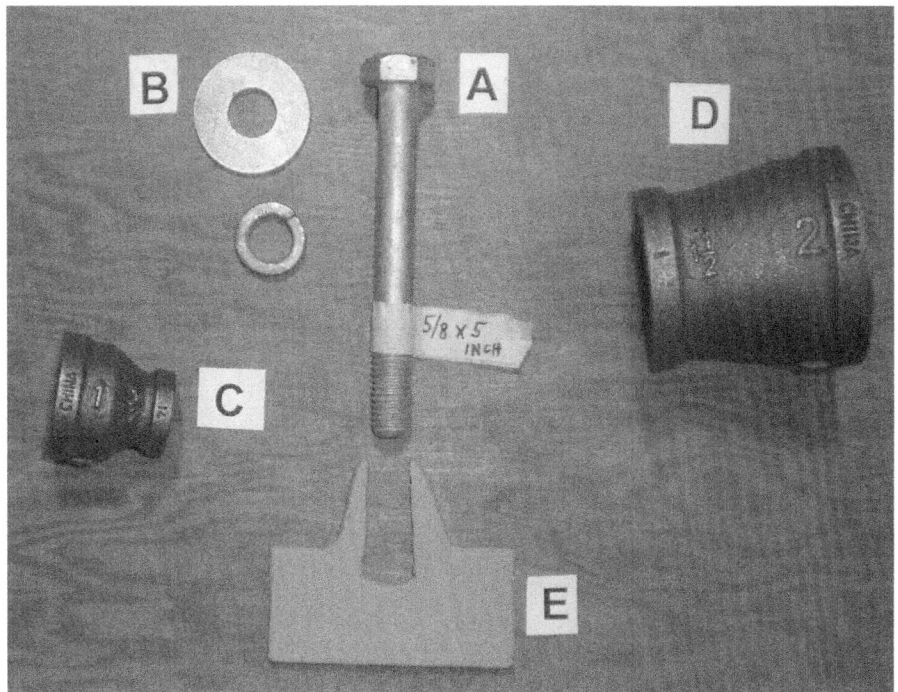

A - 5/8" x 5" bolt

B - 5/8 large washer and 5/8 lock washer (bent flat)

C - 1" to 1/2" reducer coupler

D - 2" to 1-1/2" reducer coupler

E - dart cut to shape (see dimensions below)

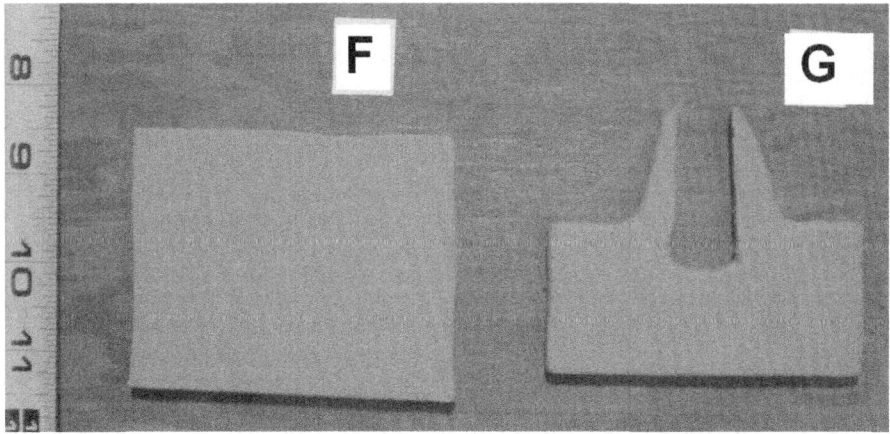

F - 1/4" thick x 4" long x 4" wide 1018 flat bar stock

G - dart cut to shape using torch, plasma cutter, or saw

A - 5/8" x 5" bolt, washer, and lock washer (bent flat)

B - 1" to 1/2" reducer coupler

C - 2" to 1-1/2" reducer coupler

D - dart cut to shape

Above: photo illustrating how the dart valve parts fit together.

Left: washers are welded to the head of the bolt. This assembly is then welded to the small reducer coupler.

Another 5/8" flattened lock washer can be welded to the small end of the reducer coupler as seen in photo.

The final product will be a solid piece that will act as an up and down (opening and closing) piston inside the main valve body.

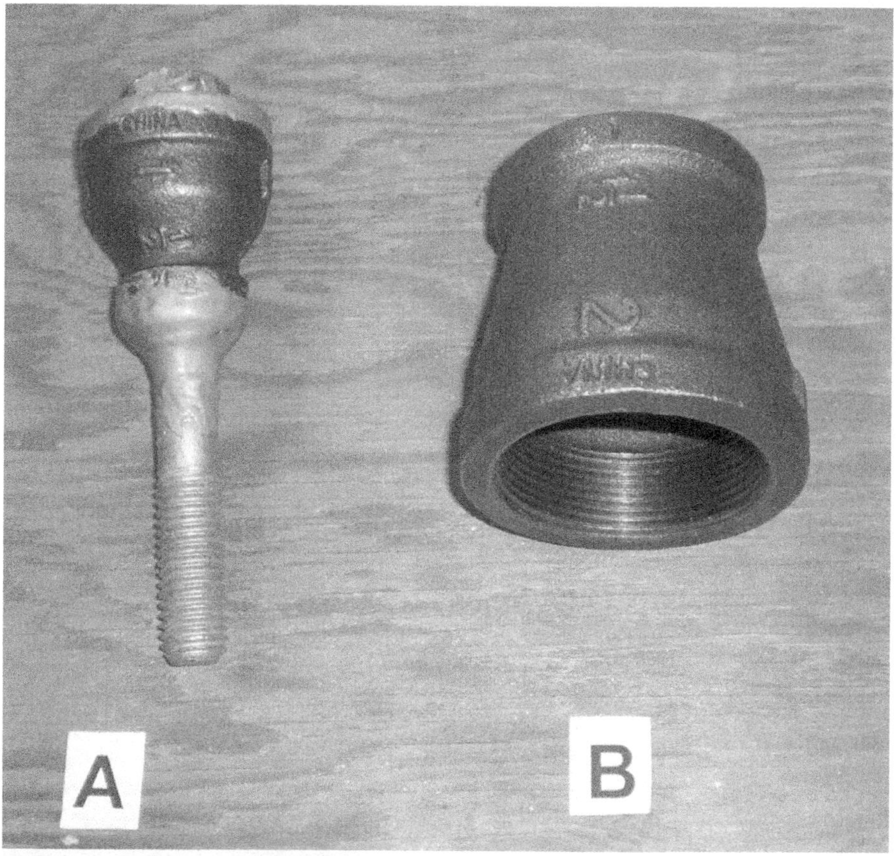

Figure A: completed valve piston shown in relation to larger reducer coupler.

Figure B: larger reducer coupler that will serve as the dart valve body. This will be directly welded into the end of the bailer tube.

Left: example of valve piston placed into it's working position inside the larger reducer coupler.

The piston valve (when finished) will move up and down a few centimeters in this manner allowing debris to enter bailer tube.

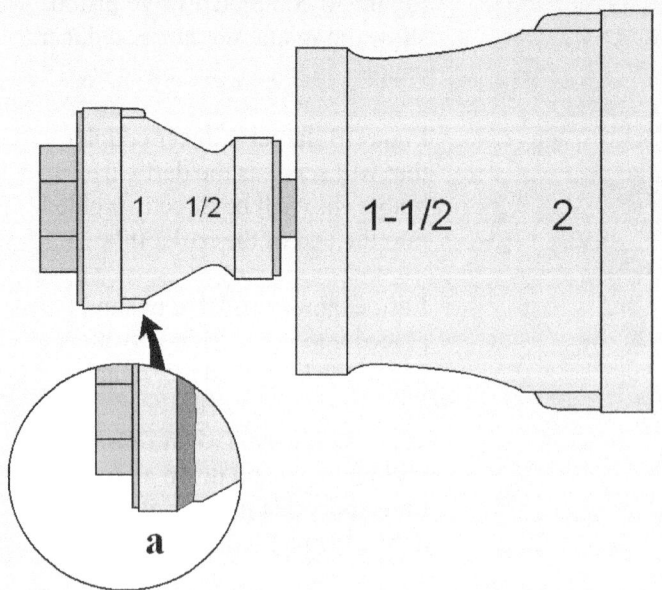

Above: tab on lip of small reducer coupler requires attention to seat properly in large reducer coupler.

Figure a: lip of small reducer coupler has been built-up using a bead of weld and ground to a bevel allowing water proof seating in large coupler.

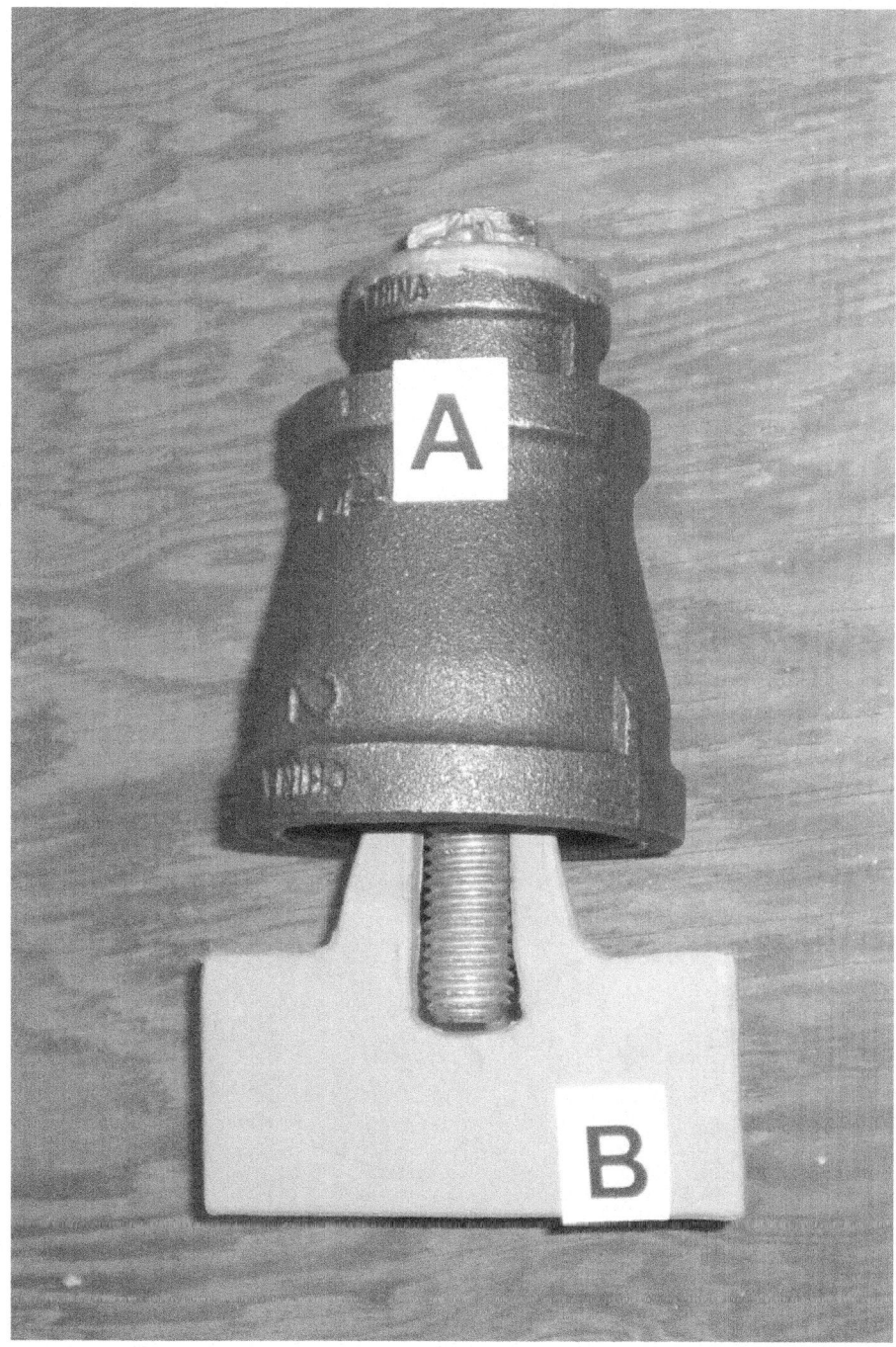

Figure A: large reducer coupler comprising the body of the dart valve assembly. Small reducer coupler and bolt assembly are seated inside.

Figure B: fabricated 1/4 inch plate steel dart welds directly to 5/8" x 5" bolt allowing a few centimeters of up and down valve movement.

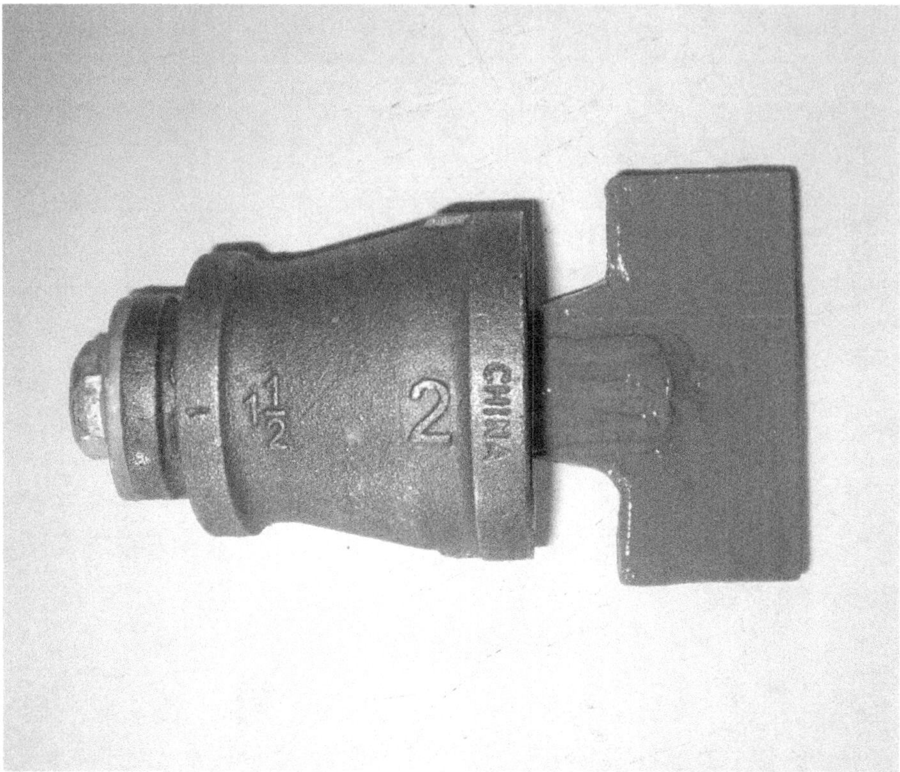
CHINA
2
1½
1

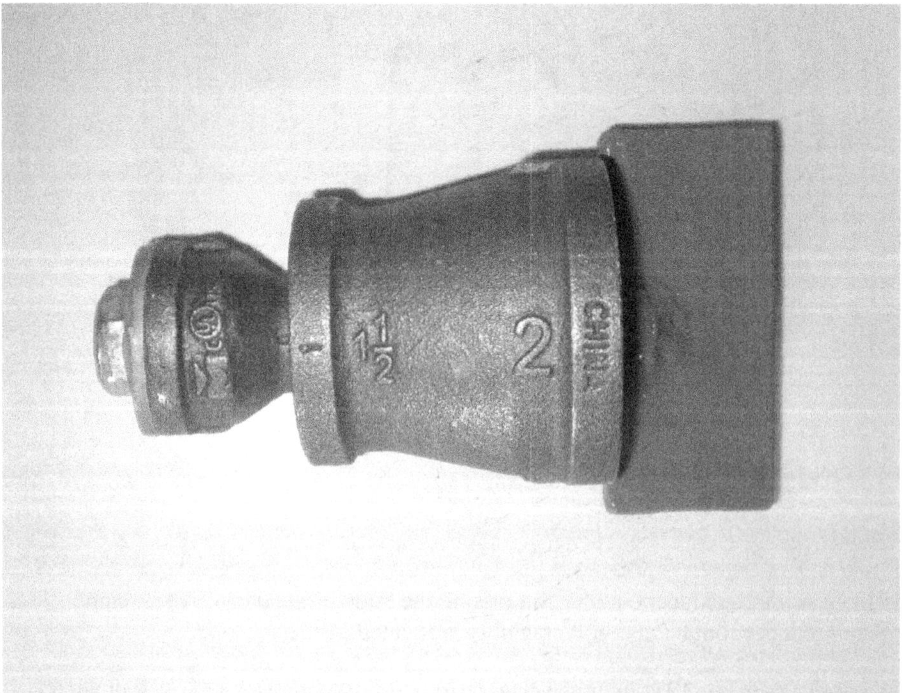
CHINA
2
1½
1

Figure A: dart welded to 5/8" x 5" bolt and valve assembly contained within larger reducer coupler body.

Figure B: larger reducer coupler body has a 2-15/16 inch wide outside diameter (OD). It contains the working valve assembly ready to be welded into bailer tube or pipe of correct inside diameter (ID).

Figure C: bailer pipe or tube used requires an inside diameter (ID) of 2-1/2 to 3 inches that will allow valve to fit and be welded in.

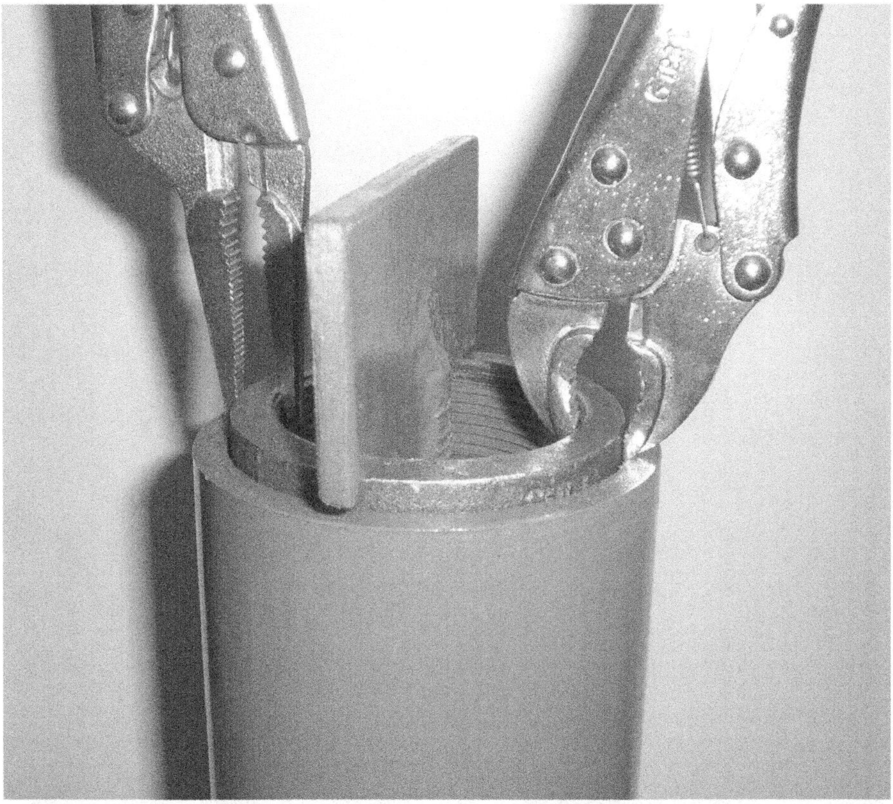

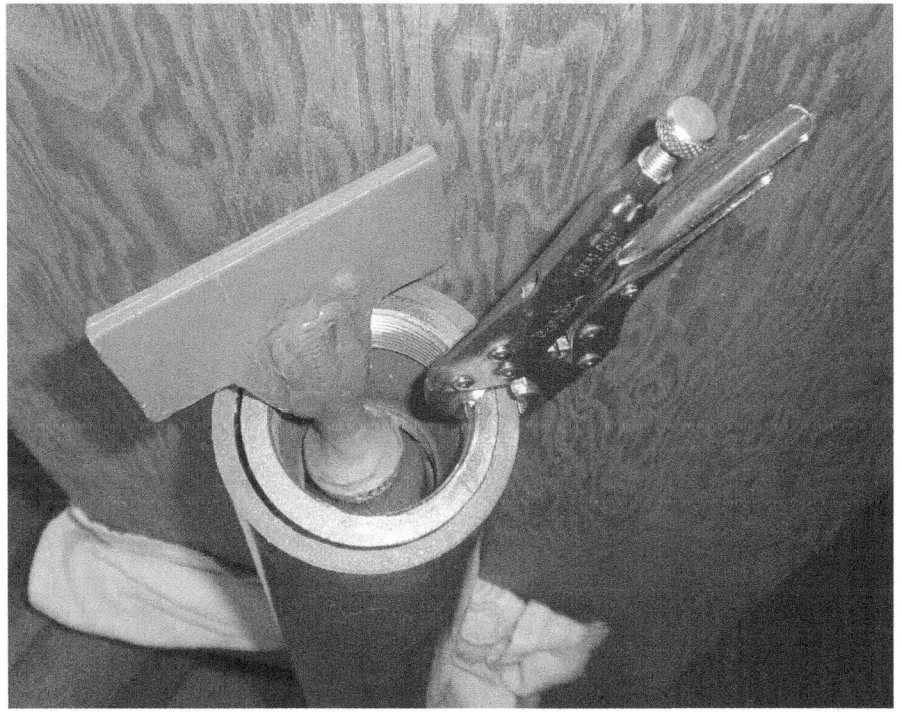

Flap Bailer

The key to fabricating a successful flap bailer is to cut the flap about 1/8 inch smaller than the inner diameter of the pipe used to house it, then form it into a slight oval. If the flap is not slightly ovaled it will not open very far. Flaps that have a 1/8 inch gap and are slightly ovaled should open into the pipe at least halfway. This will allow larger cuttings into the bailer for removal and will generate an "auto closing" effect when the bailer is withdrawn from the bore hole. Pipe edges should be ground slightly to a bevel that will allow more material in will each plunge of the bailer.

You can use a common strap hinge welded to the inside of the pipe and the flap. The gap between the flap and pipe is acceptable for removing sediment from the bore hole down to 1/8 inch in size, anything smaller (and drilling fluid as well) will experience a slow leak through this gap.

It is possible to fabricate a more leak proof flap bailer by welding additional lip material around the opening of the pipe, but care must be taken not to reduce the opening too much. The easiest solution to leaking flap valves is achieved by attaching a slightly larger diameter piece of soft (flexible) neoprene rubber to the topside of the flap. This can be done with small bolts, rivets, or a strong adhesive.

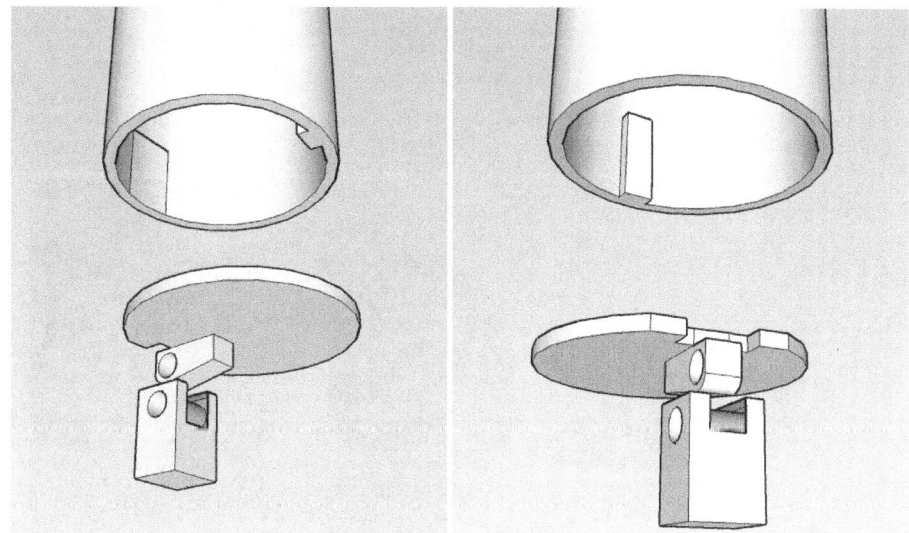

Above: flap bailer valves can be custom fabricated in a machine shop as shown, or they can be made from steel stock and common hardware store strap hinges as seen on the following pages. 1/8-1/4 inch thick plate steel is used to make the actual flap. Choose the thinnest steel that will support a full bailer tube (it's contents) during hoisting and removal from the bore hole.

Fabricating Prototype Flap Bailer

It's a good idea to build a prototype made from PVC pipe and 1/4 inch craft wood to get an idea of the flap function and range of motion inside the tube. The wood can be easily reduced uniformly (by way of a dremel tool with sanding drum) allowing for small changes and adjustments to the model flap.

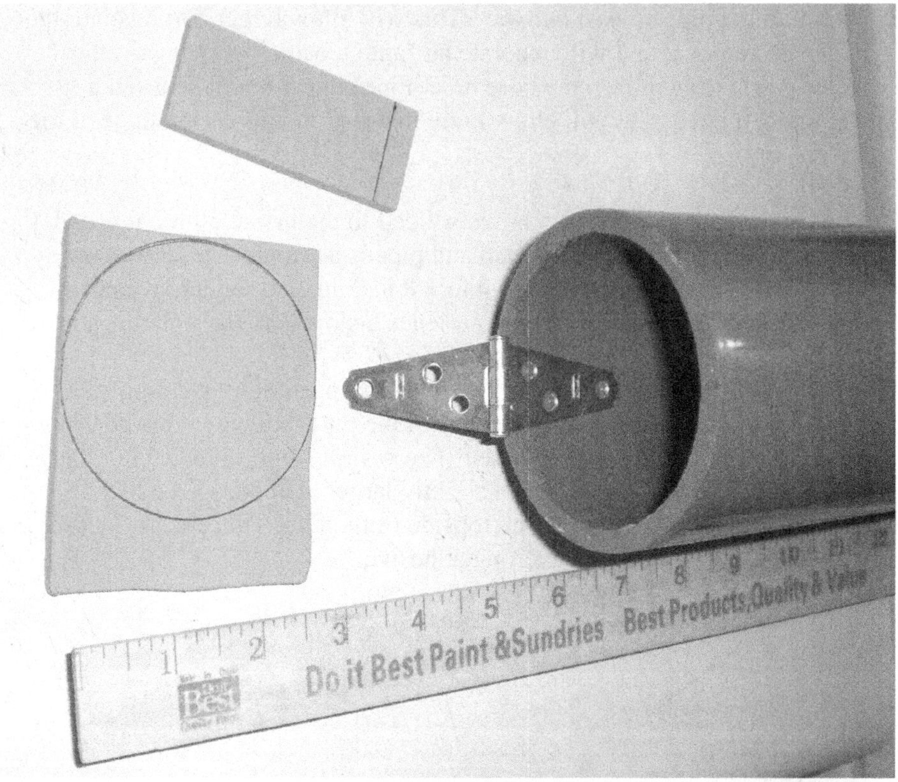

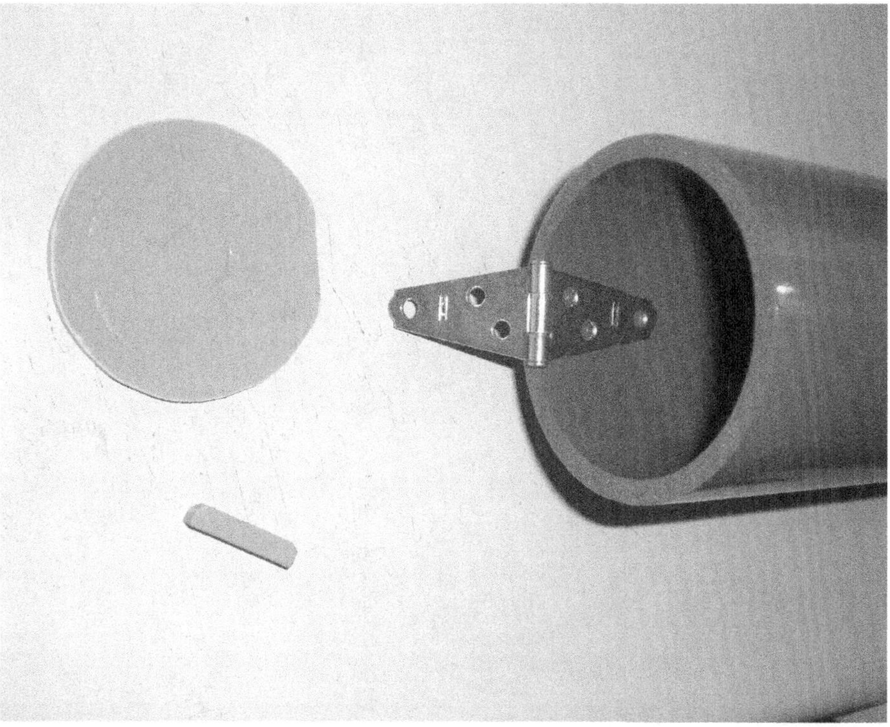

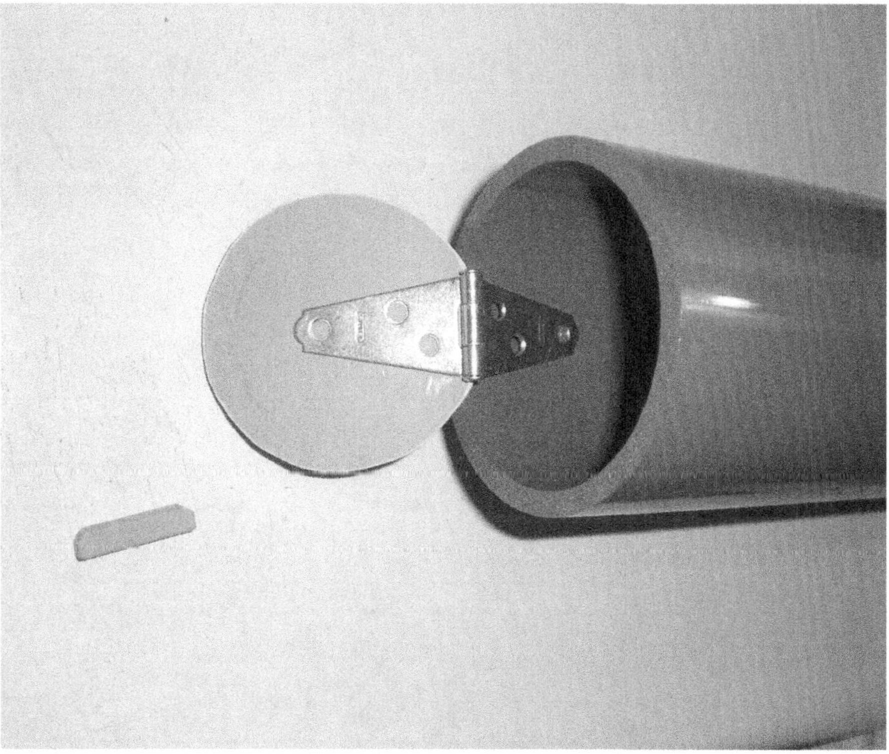

43

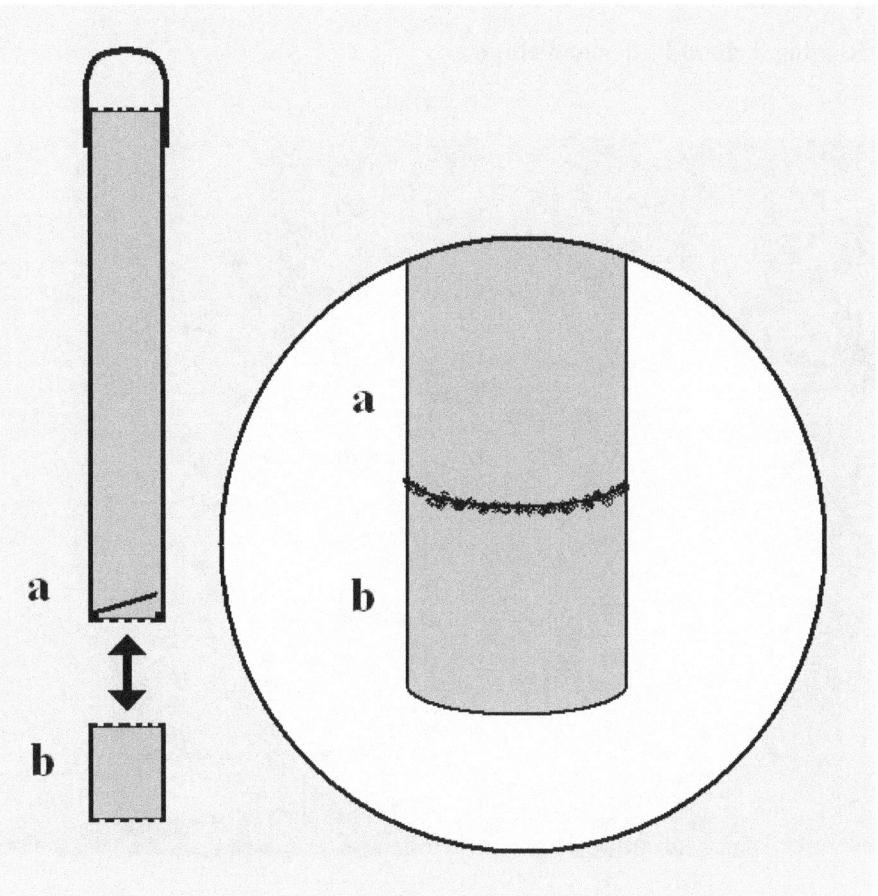

Figure a: existing bailer tube. Figure b: welded-on extension improvement.

You can weld a short extension tube to the end of the flap bailer that will prevent the flap valve from making full contact with the bottom of the bore hole. This is a useful variation that can be used to force more mud and small material into the bailer tube in a scoop fashion. It removes the drilling mud that can slow down a percussion cutting tool (allows larger cuttings to remain behind at the bottom of the hole) resulting in cleaner impacts with the cutting tool. The larger cuttings are then pulverized easier without the drilling mud in the way to slow the bits momentum. This bailer should be considered as a supplement to the regular flap or dart bailer design.

Similar additions to bailer tube ends can be designed to create a smaller diameter lip that will minimize leakage from oval shaped flap valves.

Both types should be tack welded, a method that can be ground off (if required for adjustment etc) for removal of extension tube.

Rigging Tripod Pulley and Rope

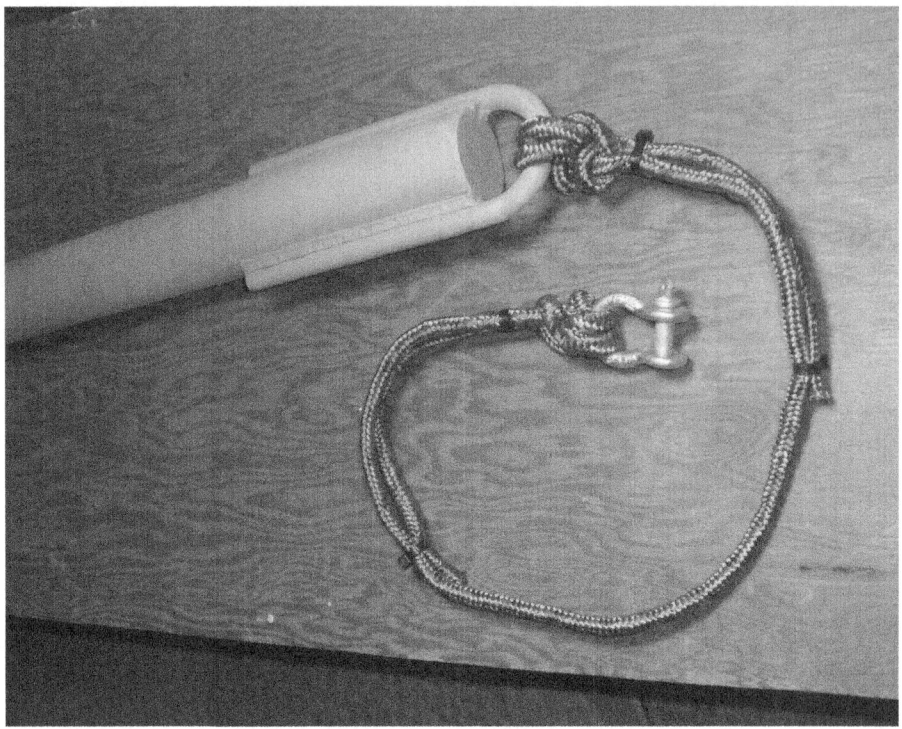

The most efficient way to rig up a percussion bit or bailer is with a one foot long rope "leader". Secure each end of the rope leader using anchor hitch knots (also known as a fisherman's the bend knot), or other strong knot. Leave one foot of rope hanging from each knot and secure in the middle of the leader cable ties. Cable ties offer a low profile way to secure loose rope ends. They can also be inspected often and easily replaced if broken or damaged. The rope leader itself should be considered disposable and replaced after any noticeable wear.

You can also use a short nylon lifting sling in place of the rope leader if available. Nylon slings will likely provide a longer service life than rope. Whatever type of leader you decide to use, make sure it is close to one foot long, perhaps two feet, but nothing as long as the bit in length as the rigging hardware weight may tend to pull it to the bottom of the bore hole perhaps snagging or becoming damaged on cutting bit blades (this is assuming that the average percussion bit is at least three feet long).

One end of the leader rope is tied directly to the bit or bailer, and the other end is tied to a 5/16 inch anchor shackle, also known as a D-ring. The rope that goes to the surface, or the working rope that is pulled by the driller, is connected to the rope leader and bit using a 3/8 inch quick-link. Rigging each percussion bit and bailer in this manner will allow for easy changes between tools without the need to constantly re-tie knots. Anchor shackles and quick links can also be connected directly to the loops on bits and bailers, but the operator should always consider a set-up that will permit rapid changes of on-line tools, from cutting bit to bailer, for example.

Above: because D-ring pins are capable of working loose, they are manufactured with a hole in the pin that allows securing (or 'mousing') the pin to the ring with wire. All anchor shackle pins should be moused and the mousing wire checked often during use. Quick links used in pairs are a better choice than anchor shackles because there is no pin to come loose.

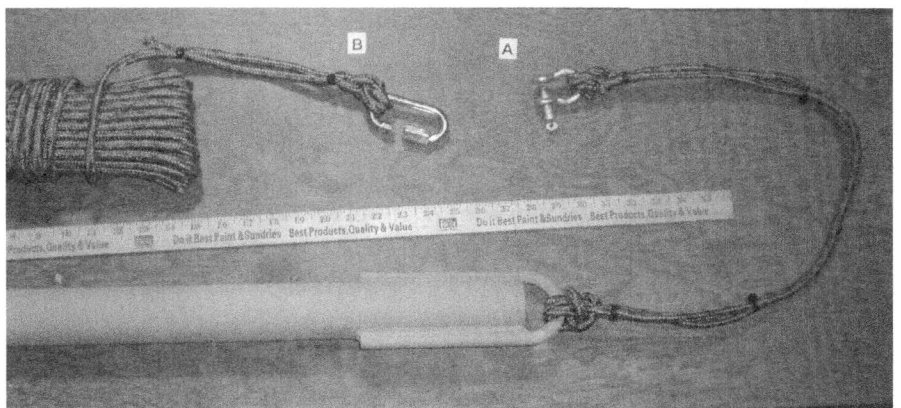

Figure A: leader connected to bit. Figure B: rope bundle and quick link.

Rope for Percussion Drill Rigging

When it comes to rigging up a percussion bit for what is likely to be several days of repeated impacts, likely thousands per day, the last thing you need is a line break. This is why it's important to choose rope carefully and make sure the rigging is 100% sound prior to commencement of work. Once a tool is lost in a bore hole drilling progress must be halted. It then becomes a retrieval operation, and if you fail to retrieve the lost tool, the bore hole must be abandoned.

Double Braid Nylon Line

Double-braid nylon line is a high-strength rope with good resistance to abrasion and easy handling. Both the core and shell of this rope is braided. The *5/8 inch diameter version of this rope has a breaking strength of 13,500 pounds. This rope is designed for heavy-duty pulley applications. It's available from marine supply stores and boating websites.

Marine Rigging Line

Pound for pound, marine rigging line is among the toughest. It offers minimal stretching and good abrasion resistance to withstand wear from pulleys and mechanical devices such as those involved in percussion drilling. Its shell is made from Spectra, a super high strength polyethylene. Its core is braided polyester. The *7/16 inch diameter version of this rope has a breaking strength of 14,300 pounds. Also available from marine supply stores and boating websites.

*Note: you can use any diameter rope that will fit within your rigging hardware, but most prefer 3/8 inch - 1/2 inch diameter for it's easier handling characteristics.

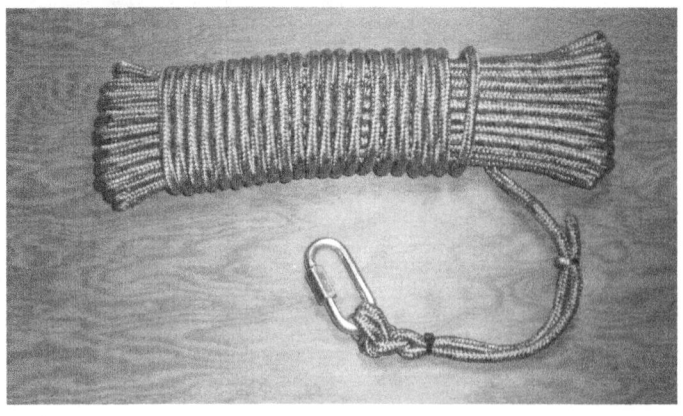

Drilling Procedure

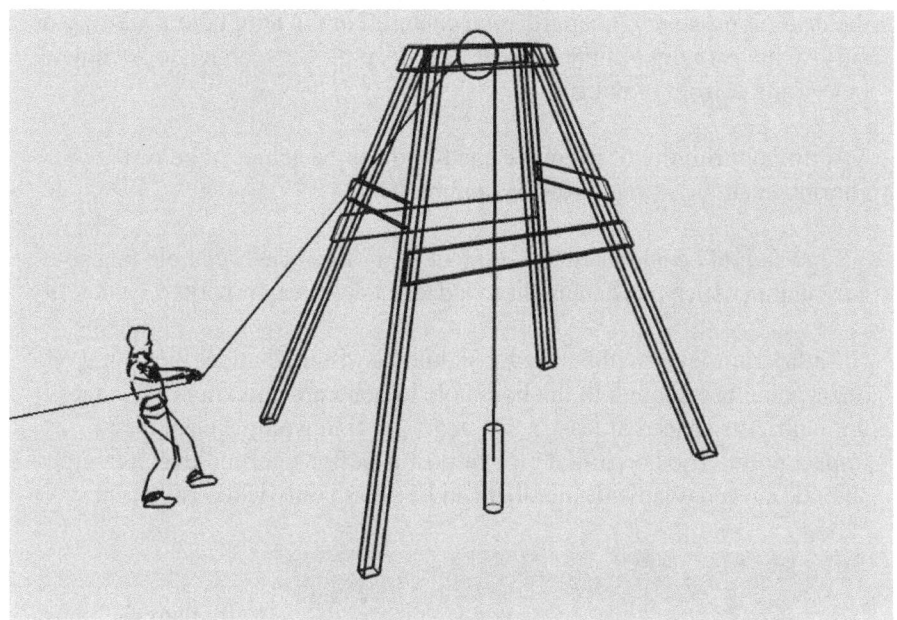

The process of operating a low tech percussion drilling rig involves two main movements: 1) manipulation of cutting bits, and 2) raising and lowering of bailers. The first goal of drilling procedure design should lie in utilizing simple physics to minimize operator fatigue and maximize tool impact during operation. Operator fatigue may not seem like a serious problem until you actually experience it. When the human engine that powers this drilling rig runs out of fuel the drilling progress is obviously halted. Speed of operation and drilling progress is the second most important goal.

By this time you will have your tools fabricated and know exactly what each weighs and what effort and rigging will be required to operate it.

Set-up Steps

1. Tripod is set-up or built onsite in the drilling location and the pulley is installed dead center at the top of the drilling rig.
2. Cutting bit is rigged and dropped into the soil to mark impact point.
3. Cutting bit is moved aside and a clam shell post hole digger or auger is used to dig as deep as possible into the ground where the impact point dictates.
4. Percussion tool drilling then resumes.

Set-Up Tips

° Tripod or quad-pod legs need to remain motionless for the duration of the drilling process to keep the rope centered in the bore hole. One way to ensure this is to dig a shallow 6-10 inch deep pit for each leg to set down in keeping it from moving.

° Additional rigidity of tripod or quad-pod can be achieved be further anchoring each leg using stakes or sand bags.

° Rope should be anchored 50-100 feet away from the bore hole in a secure manner such as attachment to a deep stake, tree, or parked vehicle.

° If a fulcrum lever will be used it should be installed under the tripod or quad-pod close enough to the bore hole to generate maximum leverage with uprights spaced at least 8 feet apart providing ample clearance of impact point. The horizontal pipe or post used for fulcrum level assembly should of a removable design that can be set-up only when called for.

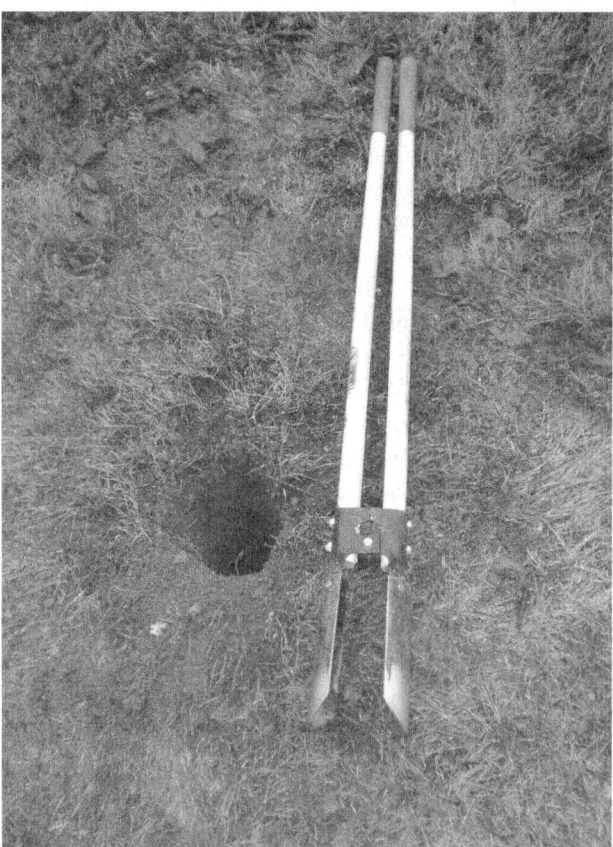

Left: clam shell post hole digger or auger is used to dig as deep as possible into the ground where the impact point of percussion tool has dictated. This depth is usually about 2-3 feet with a clam shell digger and up to 25 feet deep with an Iwan style auger.

The starter hole is not an absolute requirement, but it sets a nice pace and allows the driller to determine if an upper level temporary working casing is required to support loose soils.

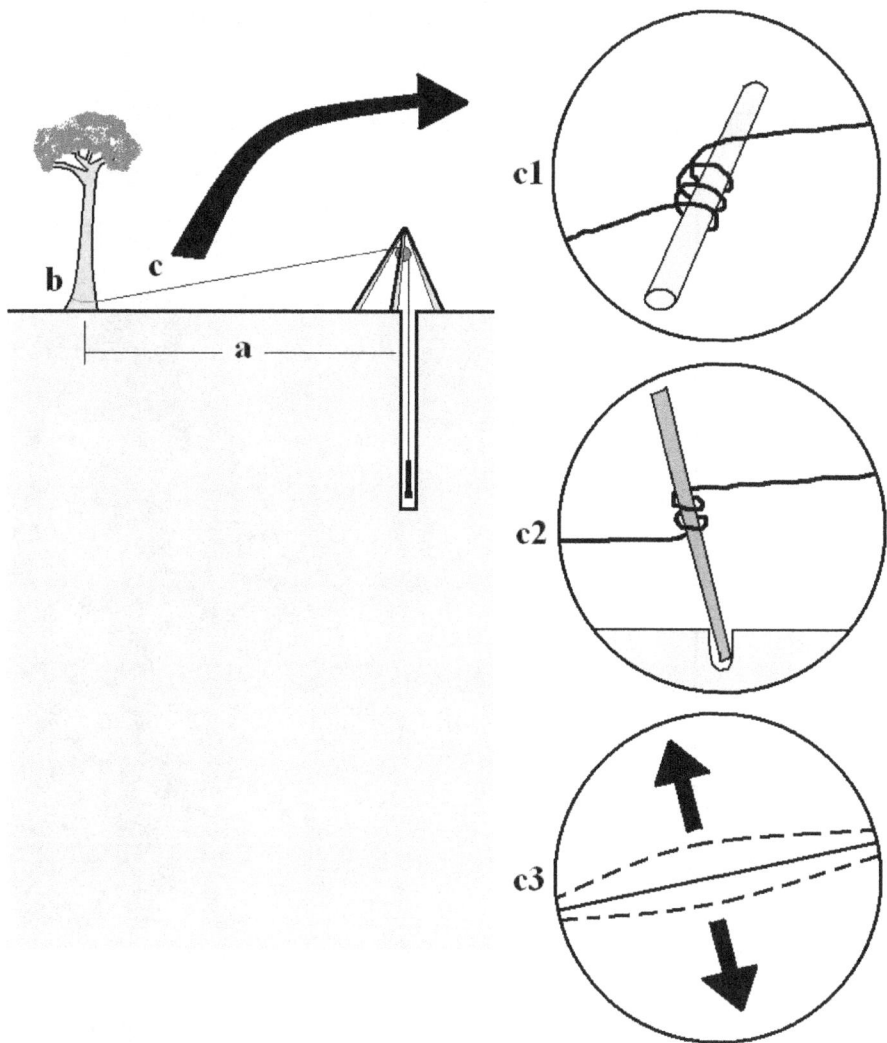

Figure a: rope length from pulley to anchor point is 50-100 feet.

Figure b: anchor points include a deep stake, tree, parked vehicle, and other secure or heavy objects.

Figure c: midway operation point where rope is at a comfortable height for user to engage using one of the methods shown below.

Figure c1: two foot long bar is looped several times through rope creating a pulling handle for short distance percussion tool strikes.

Figure c2: four foot long lever rests in a shallow pit with rope looped several times around it creating an easy pulling device for repeated tool strikes.

Figure c3: rope is moved up and down by hand to move tool.

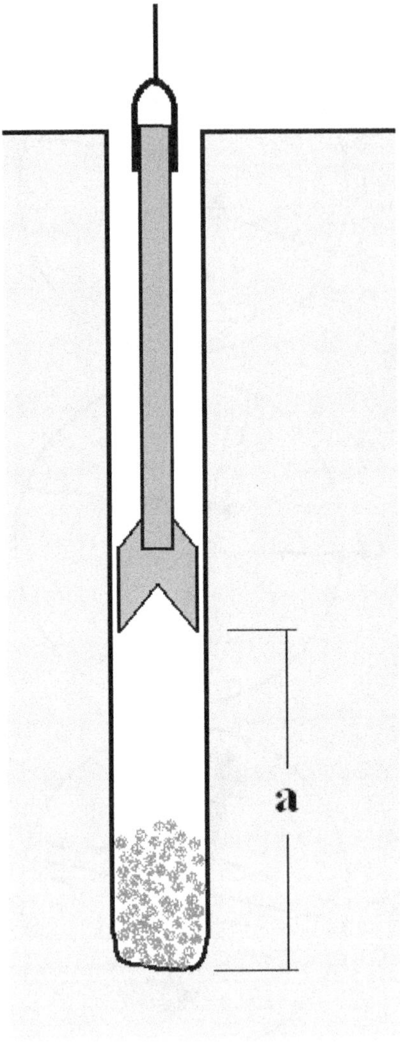

Percussion Cutting Bit Drop Distances

The distance required to drop a percussion bit and produce a sufficient cutting action is not as far as you might think. The weight of steel bits accomplishes most of the work in a 1 - 6 foot range.

Figure a:

1 - 2 feet for soft formations (rapid repeated strikes)
2 - 4 feet clays and small rocky formations (rapid repeated strikes)
4 - 6 feet larger rocks and solid formations (rapid repeated strikes)

Bentonite is a powdered clay sold in sacks that clings to bore hole walls and creates a firm reinforced layer. It's available online from well drilling supply stores. Adding a small amount of bentonite and a few gallons of water on a regular basis during drilling will form a clay wall on the inside of the bore hole preventing collapse. More of a safeguard than an absolute necessity, bentonite provides good piece of mind for not much money. Whether you use bentonite or not, water is always added to the bore hole during percussion drilling. Dumping a few buckets full of water in after each bailing creates a slurry allowing further bailing and helps float cuttings.

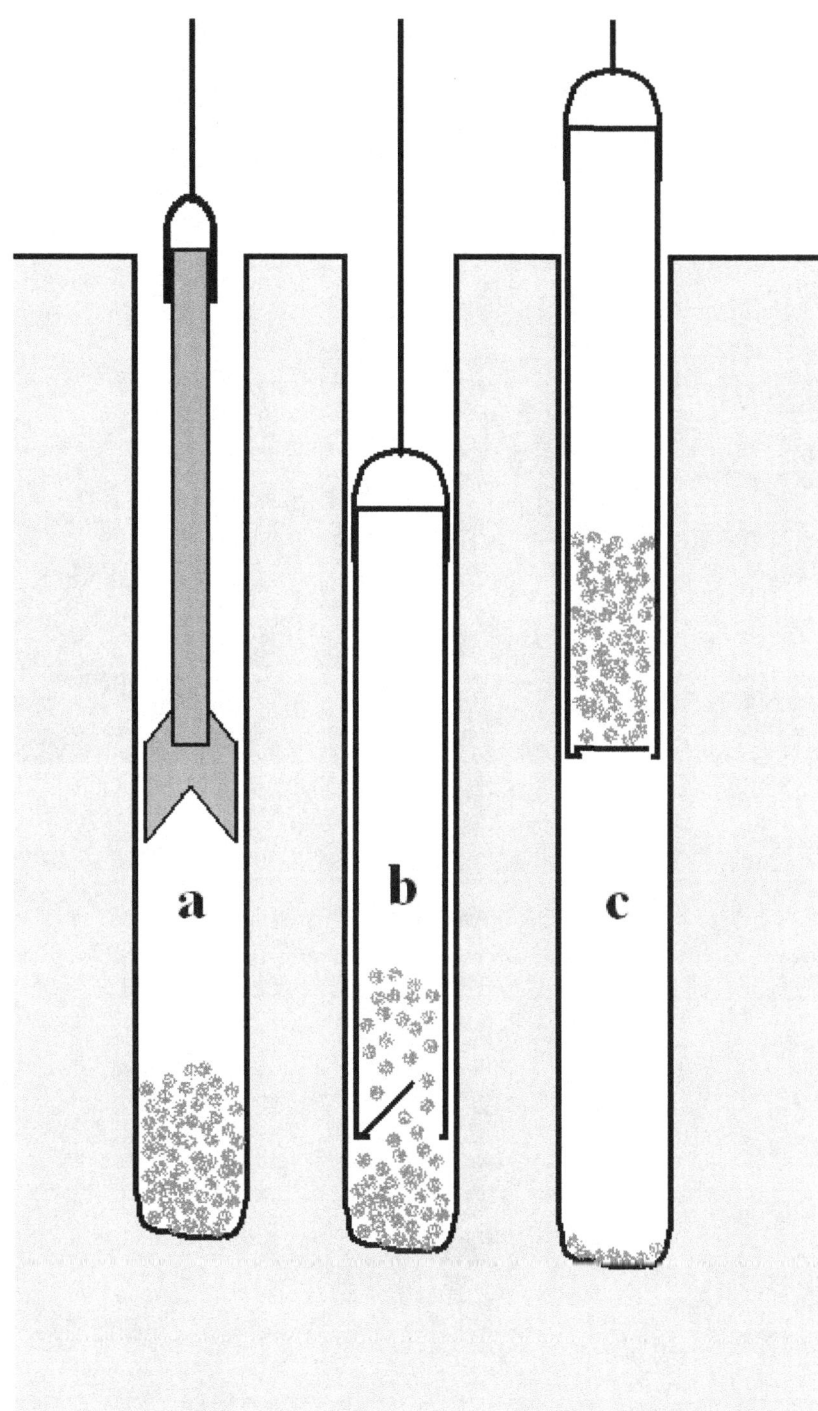

Figure a: percussion cutting bit impact generates cutting filled slurry.
Figure b: bailer is plunged into bore hole trapping slurry for removal.
Figure c: bore hole is cleared out resulting in additional depth.

Percussion drilling bit shown working in a shallow bore hole.

Top: percussion drilling bit shown working in a shallow bore hole.
Bottom: bit is removed and set aside while bailer is connected to rope.

Bailer connected to rope and lowered into the bore hole for removal of debris.

Bailer contents are emptied after removal from bore hole.

Top: bailer slurry is emptied after removal from bore hole.
Bottom: safety cover for bore hole when the work day is complete.

Sludging (Reverse Jetting) to Create a Well Hole

Overview

Another low tech well boring method often used is sludging. This is a manual style that, like percussion drilling, relies on repeated vertical percussive movement. Reverse jetting, hollow rod drilling, and hydraulic percussion are very similar to sludging, so similar in fact, that they could probably be considered the same method.

These are old techniques, perhaps renamed due to geographical location, but if one must be considered the grandfather it would likely be sludging which is an ancient Asian design originally comprised of nothing more than hollow bamboo tubes and human power. Reverse jetting, hollow rod drilling, and hydraulic percussion are essentially sludging updated with modern pipe, pipe fittings, and a check valve percussion bit.

You may have heard the term 'jetting' as a method of well hole creation. This is (obviously) the opposite of reverse jetting or sludging, but is not considered low tech because large amounts of water and pumps are required for operation. The jetting rig utilizes the powerful eroding qualities of water pumped out the end of the drilling pipe to wash ground formations away and out of the bore hole. Reverse jetting (or sludging) utilizes a small amount of water that sits in the bore hole. The drilling pipe is repeatedly impacted into the bottom of the bore hole removing ground formations that float up and are forced out of the pipe. So in essence it is a low tech reversed form of jetting that relies on human power and progresses at a much slower pace.

It is important to note early in your research of sludging that the method is only effective in light weight loose ground formations that are capable of being lifted up the drilling pipe within the fluid. The sludging effect will cease to function in any formation that is not 'sludge like'. Small rocks and mud mixed with water will create a slurry with the consistency of wet concrete rather than sludge, causing failure.

Sand, soft silt, light clays, etc are acceptable formations to try sludging in. The sludging plan presented here has the added benefit of utilizing a standard 1-1/4 inch galvanized pipe assembly as the main drilling pipe. This gives drillers the option of using a regular PVC casing when finished, or connecting a standard 1-1/4 inch drive point (screen) to the pipe for achieving slightly deeper depths and keeping the pipe in-hole.

Since these methods are designed for soft formations, possible collapse of the bore hole is an issue. Even though they rely on the bore hole always being filled with water (the water produces enough hydraulic pressure to hold the side walls of the bore hole open in most cases), the additional measure of a temporary casing for the top 10 feet may be required. Also, adding bentonite clay to the drilling fluid (because it reinforces bore hole walls) is always a good idea.

Most wells created using these methods are cased with 1-1/4 to 2 inch diameter PVC pipe, and most drilling rigs are made from 1-1/4 inch galvanized pipe and fittings. If conditions are right, this method can reach depths of over 100 feet deep.

Low tech sludging has been used for centuries in Asia (with bamboo tubes and the operators hand as a valve), but a better version that utilizes common hardware store parts has been developed for use by Terry Waller of the organization 'Water for All International'.

waterforallinternational.org

The original sludging idea, along with modern pipe and hardware store bit, is commonly referred to as 'The Baptist Method' of sludging. It's an improvised version of hollow rod drilling that utilizes a rather ingenious cutting-bit-check-valve combination tip made from pipe reducers, couplers, and a bolt. It threads directly to a length of galvanized pipe.

You can fabricate a large selection of these style bits for varied applications. Using galvanized pipe couplers and reducers as the foundation, you can come up with lots of variations expanding on the original idea.

These bits can be welded together in a small metal working shop equipped with such tools as a vice, welder, and angle grinder.

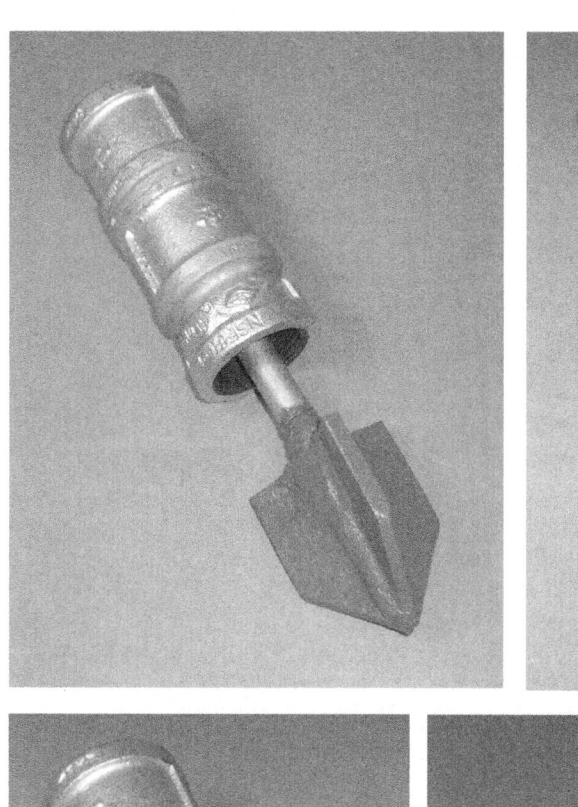
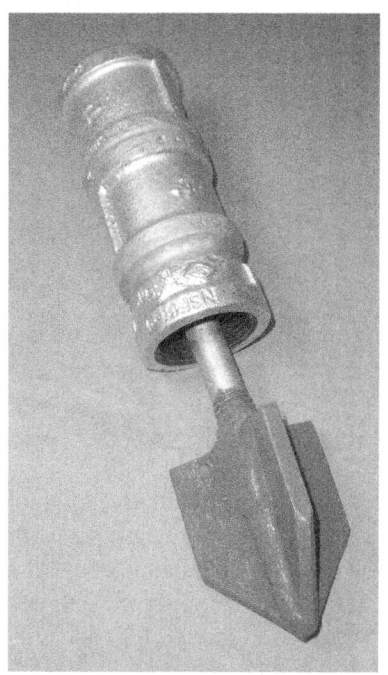
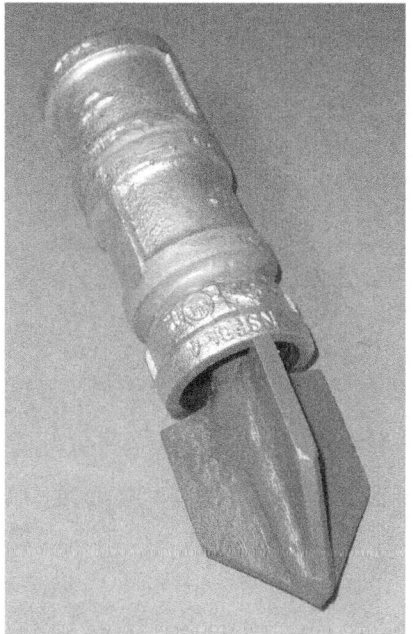

The Terry Waller designed sludging bit features the benefit of utilizing common hardware store parts such as pipe couplers, reducer couplers, and bolts. Cutting and welding of 1/4 inch plate steel blades is required and can be preformed by the do-it-yourself home builder or a local welding shop.

Do-It-Yourself Sludging Rig Components

Terry Waller Design Sludging Bit with Built in Valve

(2) 1-1/4 inch galvanized steel pipe coupler (per bit)
(1) 1-1/4 to 3/4 inch galvanized steel pipe reducer coupler (per bit)
(1) 5/8 x 5 inch steel bolt (per bit)
(1) 1/4 inch thick x 2 inch wide flat bar stock

Note: black pipe fittings can be used in place of or interchanged with galvanized fittings if required.

Sludging Pipe Assembly

(1) 1-1/4 inch 90 degree galvanized elbow
(1) 1-1/4 inch 45 degree galvanized elbow
(4) 1-1/4 inch galvanized steel pipe couplers
(1) 1-1/4 inch x 10 inch galvanized nipple
(4) 1-1/4 inch x 10 foot schedule 40 galvanized pipes (23 lbs per 10 foot)

Note: the number of pipes and couplers required depends on depth of bore hole desired, enough pipe for a 40 foot well is shown.

Note: the 1-1/4 inch galvanized pipe weighs 2.27 lbs per foot.

The sludging pipe assembly is heavy to begin with and becomes heavier as you add pipe. An essential part of the sludging rig is a fulcrum lever set-up as described in the previous chapter 'Percussion Drilling to Create a Well Hole'. The lever is tied to the sludging pipe by rope generating the rapid up and down piston action required to penetrate ground formations. This can be supplemented by erecting a tripod and pulley operated rope over the bore hole that can add lifting power when operated by additional crew member.

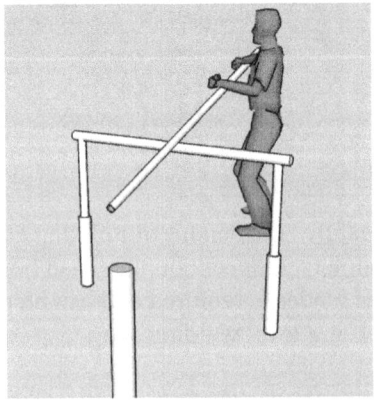

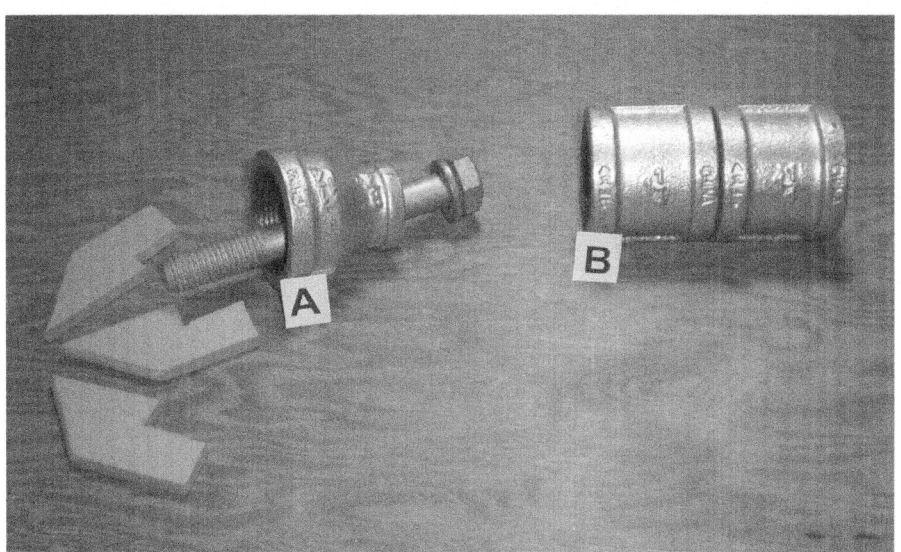

Top Figure A: 1-1/4 to 3/4 inch galvanized steel pipe reducer coupler, 5/8 x 5 inch steel bolt, and 1/4 inch thick fabricated dart blades. Dart blades can be cut using a torch, plasma cutter, or saw and should have the cutting edges beveled using a grinder.

Top Figure B: (2) 1-1/4 inch pipe couplers are welded together making up the valve chamber and the threaded attachment point for sludging pipe.

Bottom Photo: couplers and reducer coupler welded together with the bolt placed in it's operating position.

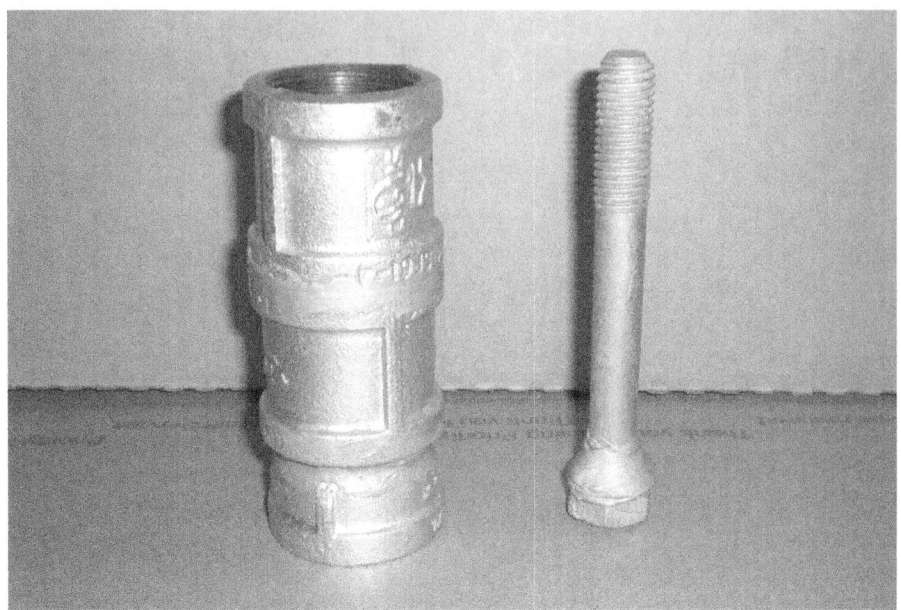

Figure A: 5/8 inch lock washer (that has been flattened) is welded to the bolt. Additional weld is then applied to the neck of the 5/8 x 5 inch bolt and washer creating a steel formation that is then shaped by angle grinder creating a bevel that will seat into the 3/4 inch opening of the reducer acting as a valve.

Figure B: inside view of 1-1/4 to 3/4 inch galvanized steel pipe reducer coupler after it has been welded into the two 1-1/4 inch galvanized steel pipe couplers. This is the regulated opening that will allow debris into the pipe during the down stroke of the drilling process.

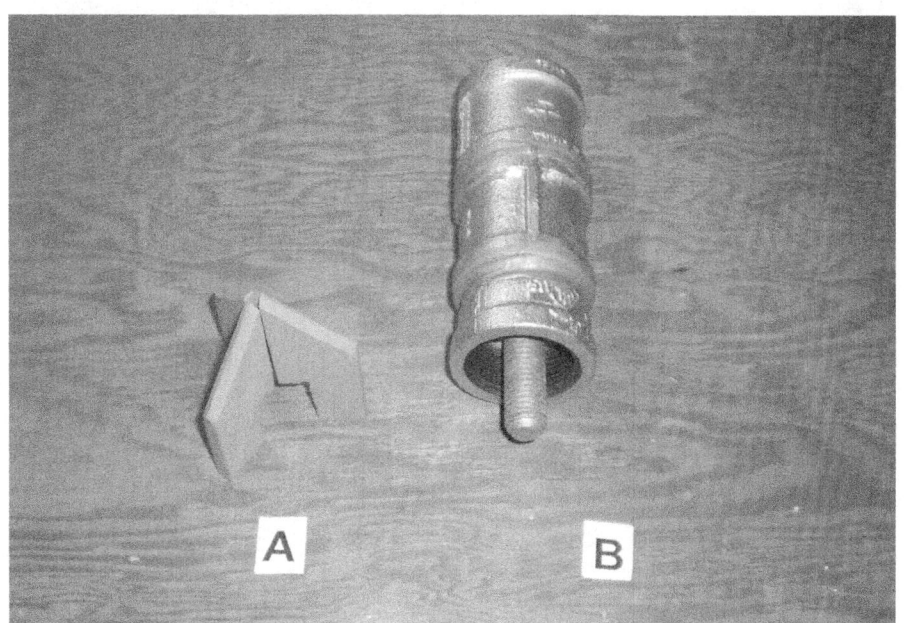
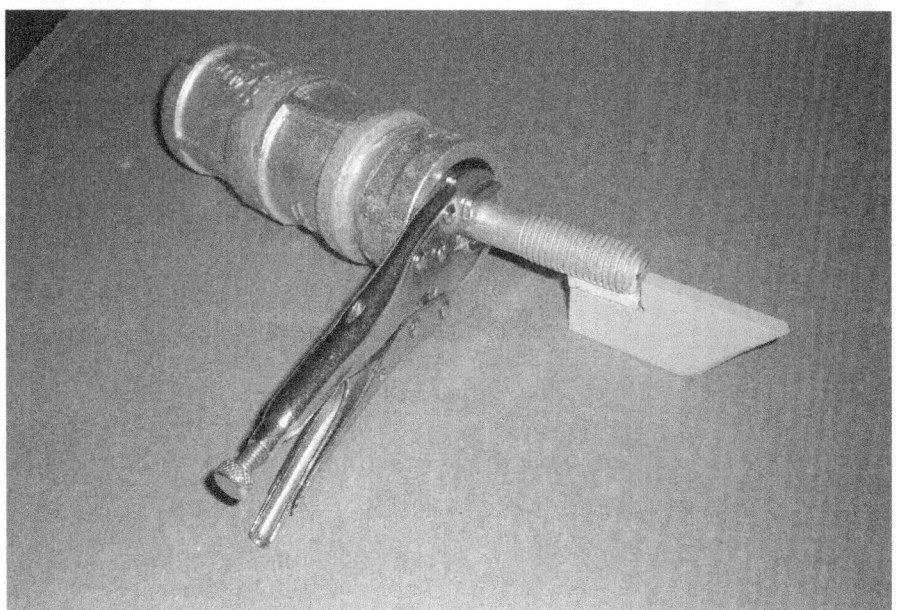

Figure A: dart blades arranged in mounting position ready for welding.

Figure B: bit assembly and bolt prior to dart blade installation.

Bottom Photo: bolt is secured with vice grips or a bench mounted vice prior to the positioning and welding of dart blade segments.

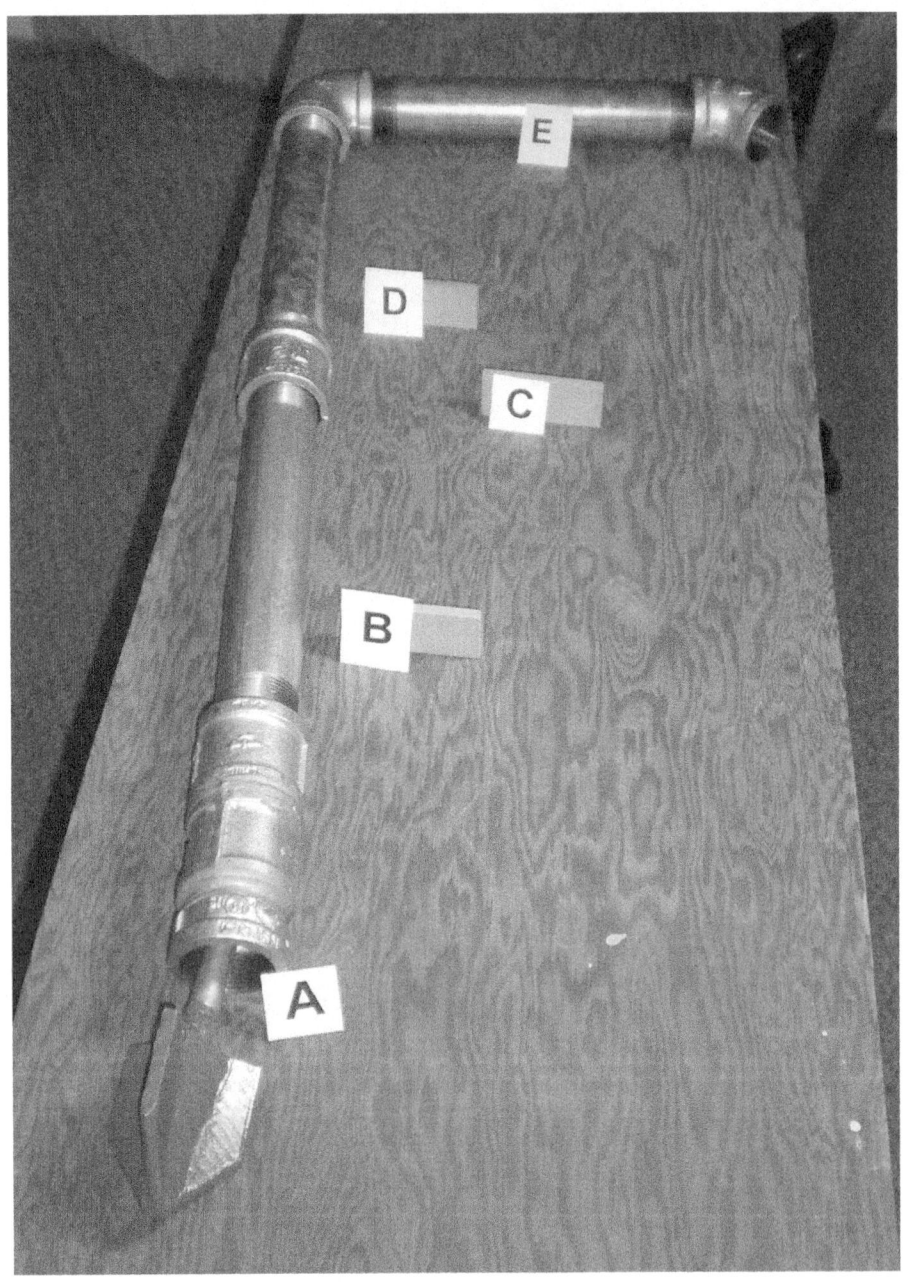

Figure A: pre-fabricated sludging bit.
Figure B: 1-1/4 inch galvanized pipe (shown short for example only).
Figure C: 1-1/4 inch pipe coupler.
Figure D: 1-1/4 inch galvanized pipe (shown short for example only).
Figure E: top of sludging pipe assembly comprised of a 1-1/4 inch 90 degree elbow, 1-1/4 inch x 10 inch pipe nipple, and a 1-1/4 inch 45 degree elbow. Lever rope is attached around the 90 degree coupler.

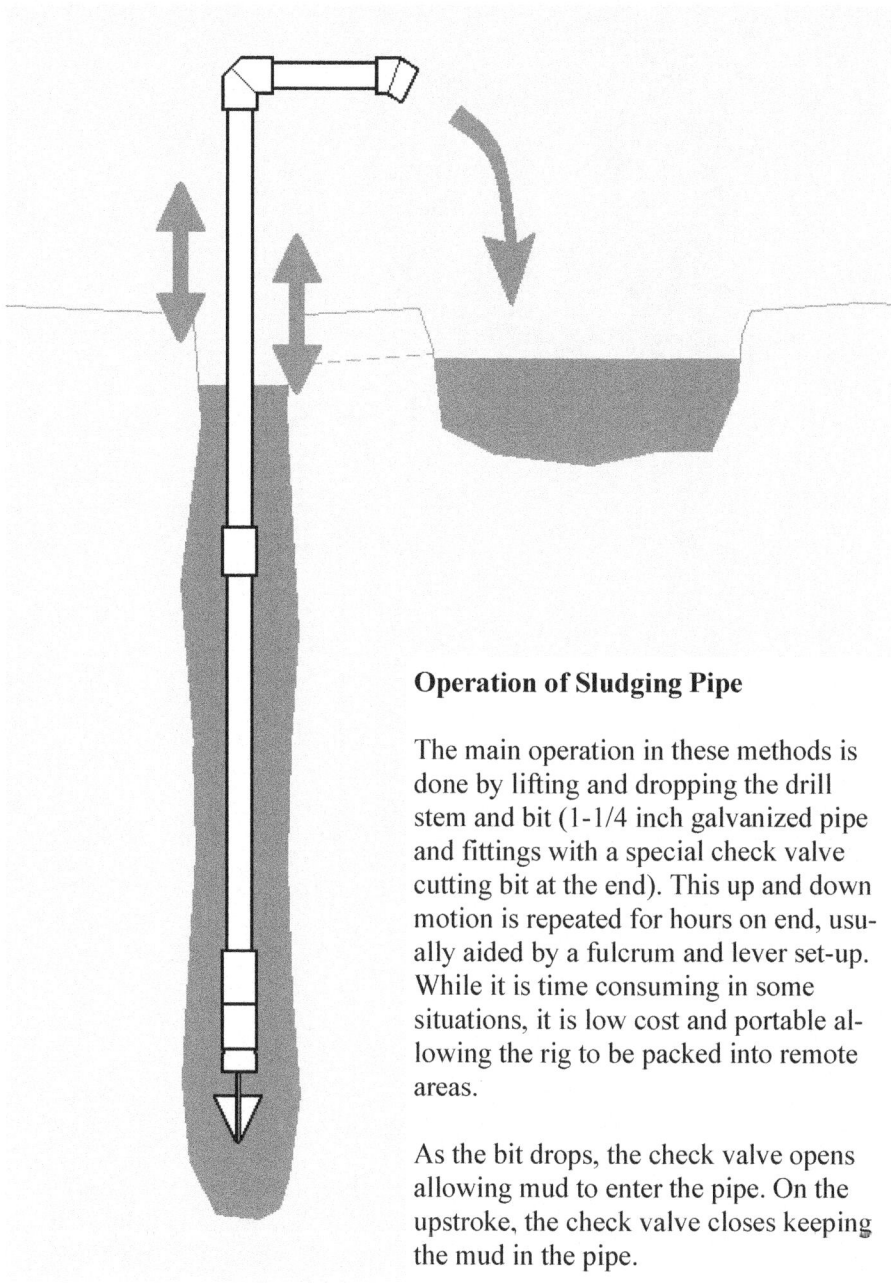

Operation of Sludging Pipe

The main operation in these methods is done by lifting and dropping the drill stem and bit (1-1/4 inch galvanized pipe and fittings with a special check valve cutting bit at the end). This up and down motion is repeated for hours on end, usually aided by a fulcrum and lever set-up. While it is time consuming in some situations, it is low cost and portable allowing the rig to be packed into remote areas.

As the bit drops, the check valve opens allowing mud to enter the pipe. On the upstroke, the check valve closes keeping the mud in the pipe.

Eventually the pipe fills and the slurry is discharged into the mud pit at the surface near the bore hole. This mud pit traps removed debris while allowing useable water to flow back into the bore hole via a shallow trench. This recycling action allows for successful sludging with only a limited amount of water.

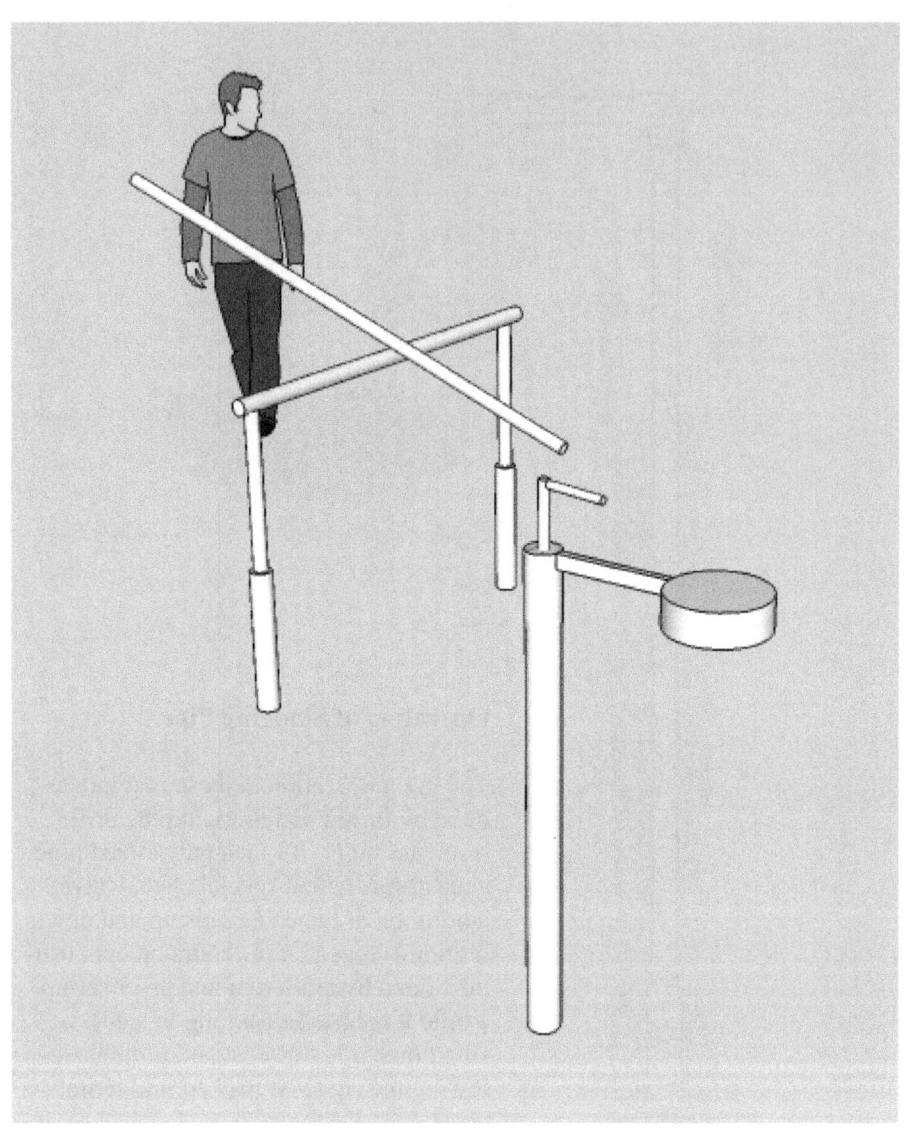

Precautions to be observed when sludging

The fluid in the borehole produces pressure that keeps it open and resistant to collapse. Sludging-drilled boreholes must be kept full of fluid during the entire drilling and well installation process. Obviously, if you attempt to utilize sludging in porous ground that will not contain your drilling fluid (such as gravel and certain sands), the borehole will self-drain resulting in collapse. If self-draining ground formations are only encountered in the first few feet, with a firm fluid-containing formation laying below, a temporary casing 2-3 feet long can be installed to hold back the loose unstable upper walls.

Hand Auger to Create a Well Hole

Overview

Use of an auger in boring a well hole is the sole method in this book that does not rely on vertical movement and vertical operating apparatuses. Instead, the cutting, or auger head, rotates, cuts, and penetrates into the ground in a rotary drilling fashion.

Depths of up to 50 feet can be reached with the hand auger, but depths of 20-30 feet are more common. Augers might often be mentioned in the same breath as shallow wells, but a more diplomatic term might be 'medium depths' wells. This is considering the fact that a shallow well is anything under 25 feet, and the hand auger is easily capable of creating a 30 foot well with a 4 inch casing and submersible pump mounted inside.

Turning the auger is the easy part of the process. Lifting the full loaded assembly out of the bore hole often is what generates operator fatigue. Also, as drilling progresses, auger assemblies become heavier from the steel extensions or pipes. To assist with lifting, two pipe wrenches can be used (as lifting devices), or a tripod with winch can be used to lift heavy auger and extension assemblies out of the hole.

Each load of material removed from the auger should be examined and recorded in the well record under the depth it was found. The water table is usually reached when the material in the bore hole becomes naturally muddy and no longer clings to the auger for removal. At this point a bailer can be used to remove mud and cuttings.

Continued augering through the water table becomes increasingly difficult. This is caused by the sticky mud at the bottom of the bore hole creating more cohesion, plus the extra water weight in removed materials. Boring at least 5 feet into the water table is recommended, but you can only go as far as you are able. The further into the water table you can successfully bore, the higher your rate of well recharge (water) will be.

Augers can also be used to drill the upper portion of a well that is completed with the cable tool method or a drive point installation. Augers tend to fail when large rocks or solid formations are encountered, but you can save augered bore holes by switching to percussion drilling when such obstacles are encountered.

Fabrication

Using an auger to bore a well hole is simplicity itself, yet the full auger sets so common in the past are difficult to find today. These would traditionally come with several bits including a regular auger, spiral auger blades of varying diameter, extension rods, and a t-bar handle. There are a few stores on the internet that specialize in ground sampling tools that still sell them, but finding a full auger set for sale locally is difficult.

Building your own steel auger set is possible with a minimum of shop equipment, or it can be done on a for hire basis by a local metal fabrication shop.

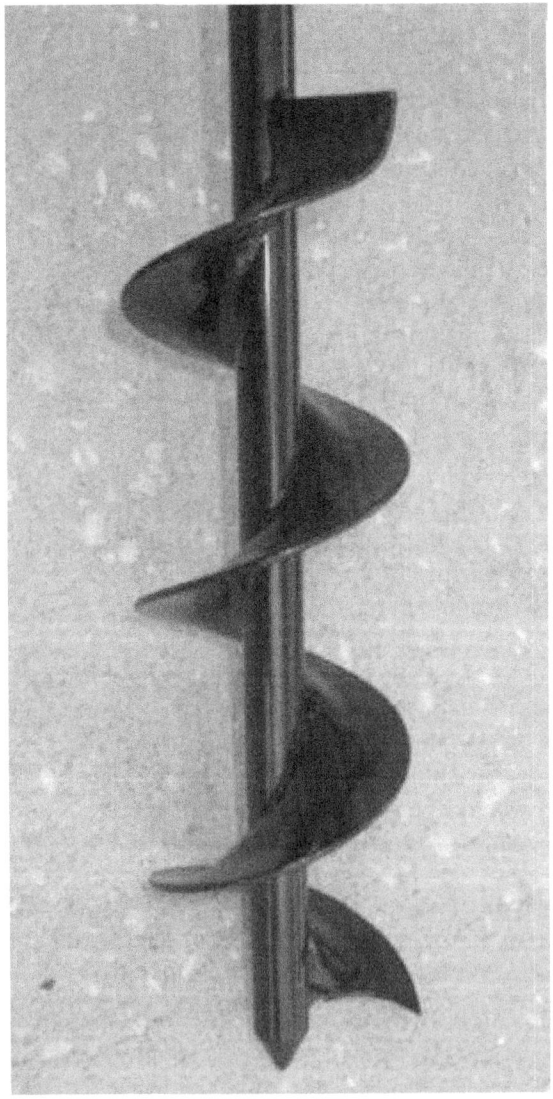

The most direct process for making a simple auger is to cut discs out of 1/8 inch plate steel. Cut discs to desired typical auger diameter of 2-8 inches wide. Cut a hole in the center of each disc and make a cut from the hole to the outside of the disc. Use a vice or press to bend the discs slightly so that the discs can be welded together at the cut line. Place the discs over a shaft and stretch the discs out, as the discs stretch out they will tilt over and the hole will jamb on the shaft. The size of the internal hole will determine the pitch of the auger. Larger holes will allow a steeper pitch of the blades. Once the auger is fully stretched, weld the joints and the flights to the shaft. You can use various diameter discs, shafts, and holes to create augers and spirals for different ground conditions.

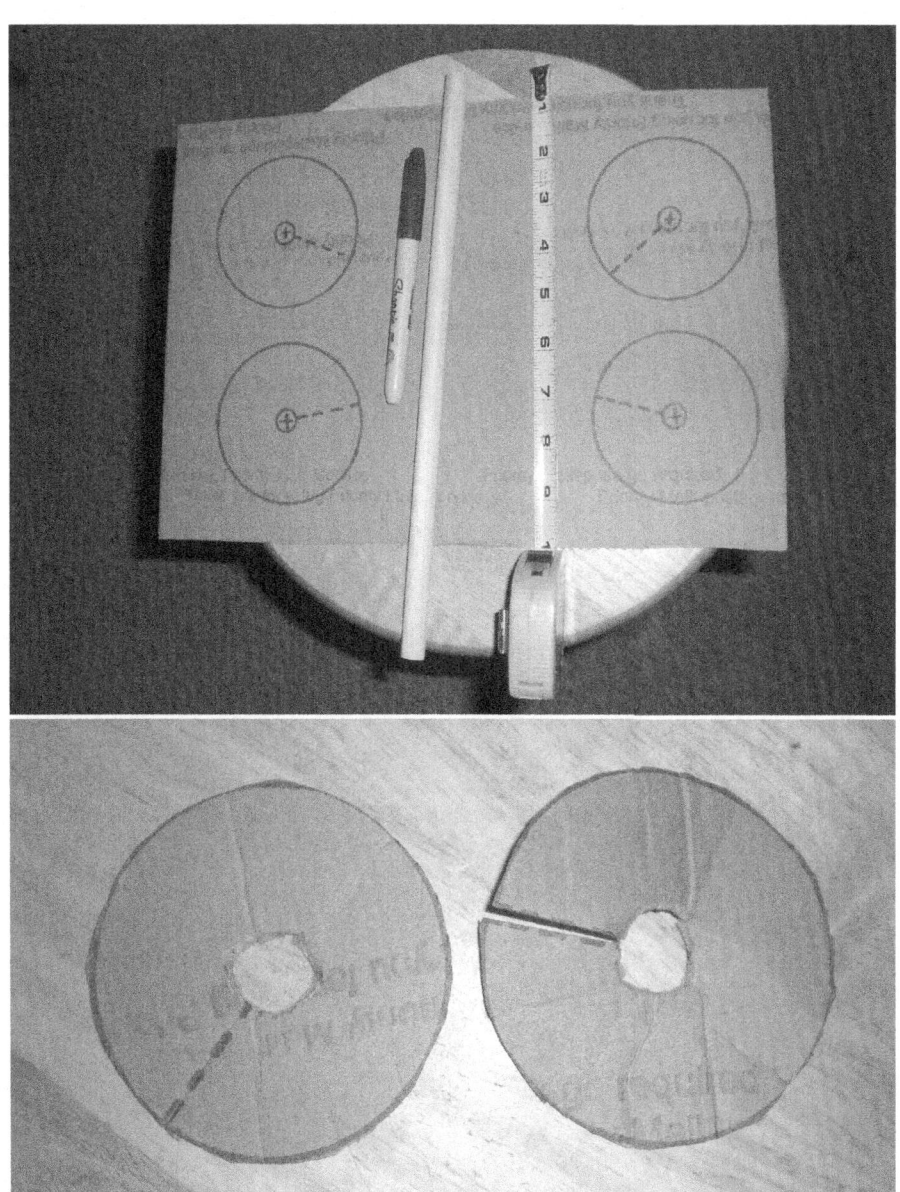

Building Auger Blade Prototypes From Cardboard Discs

Before fabricating with steel, it's often a good idea to gain a feel for the way spiral auger blades are made by building a simple cardboard prototype model. It takes only a few minutes to trace a circle around a 3-4 inch diameter can or bowl, cut out the circles, and form them into auger blades for attachment to a 1/2 inch wood dowel. Make sure to cut the center holes about an inch in diameter to allow for bending into a spiral over the 1/2 inch dowel. Four discs will create a spiral auger blade about one foot long.

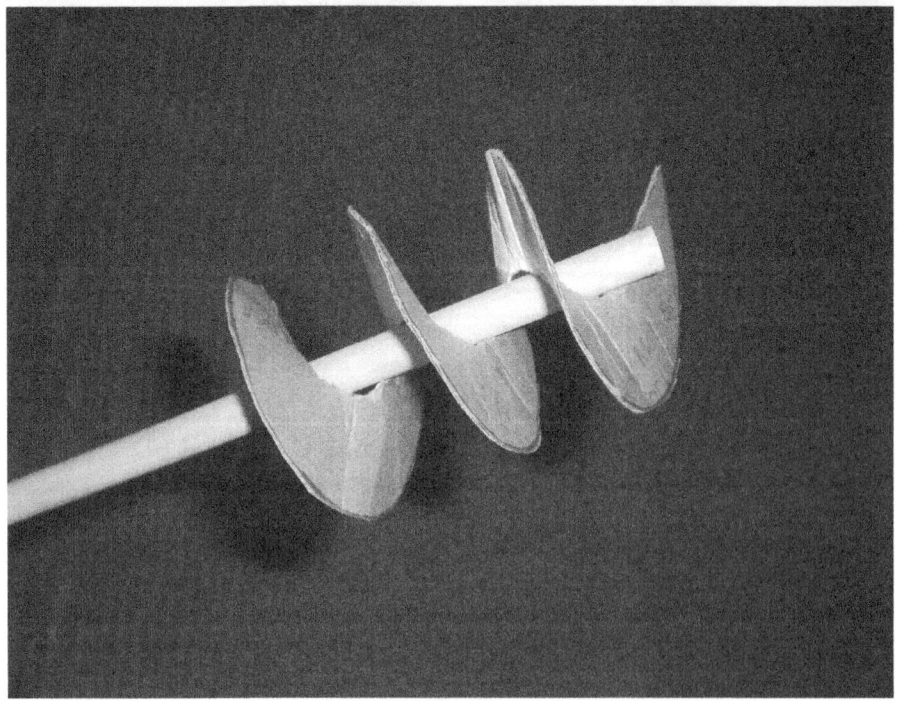

Top: two disc spiral auger blade. Bottom: three disc spiral auger blade.

Iwan Brothers Company "The Iwan Auger"

The Iwan Auger was a patented design sold for many years, and amazingly, is still available for purchase today under the Iwan name. The Iwan Auger remains as a design that is manufactured by a variety of companies. Seymour Mfg. produces several variations of the Iwan Auger.

To search online use keywords: *Seymour Iwan Auger*

Iwan augers usually have a fixed 4-8 inch diameter head (adjustable head models are also available) with a 21 inch handle and an overall length of 46 inches.

The beauty of this design is a 3/4 inch pipe fitting socket that accepts standard 3/4 inch galvanized pipe for use as extensions. If you need a deeper hole, just add a section of 3/4 inch pipe and keep on digging.

This makes it an ideal tool for boring shallow well holes up to 25 feet deep. It works best in sand and clay. Large rocks will stop progress all together but the Iwan Auger head is able to trap and extract smaller (1-2 inch diameter) rocks.

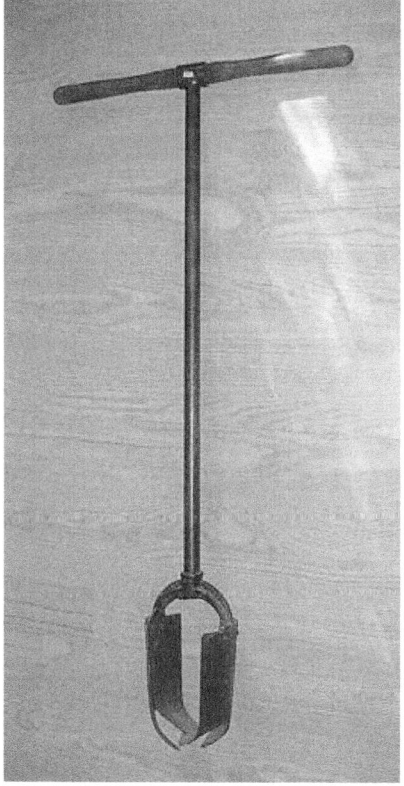

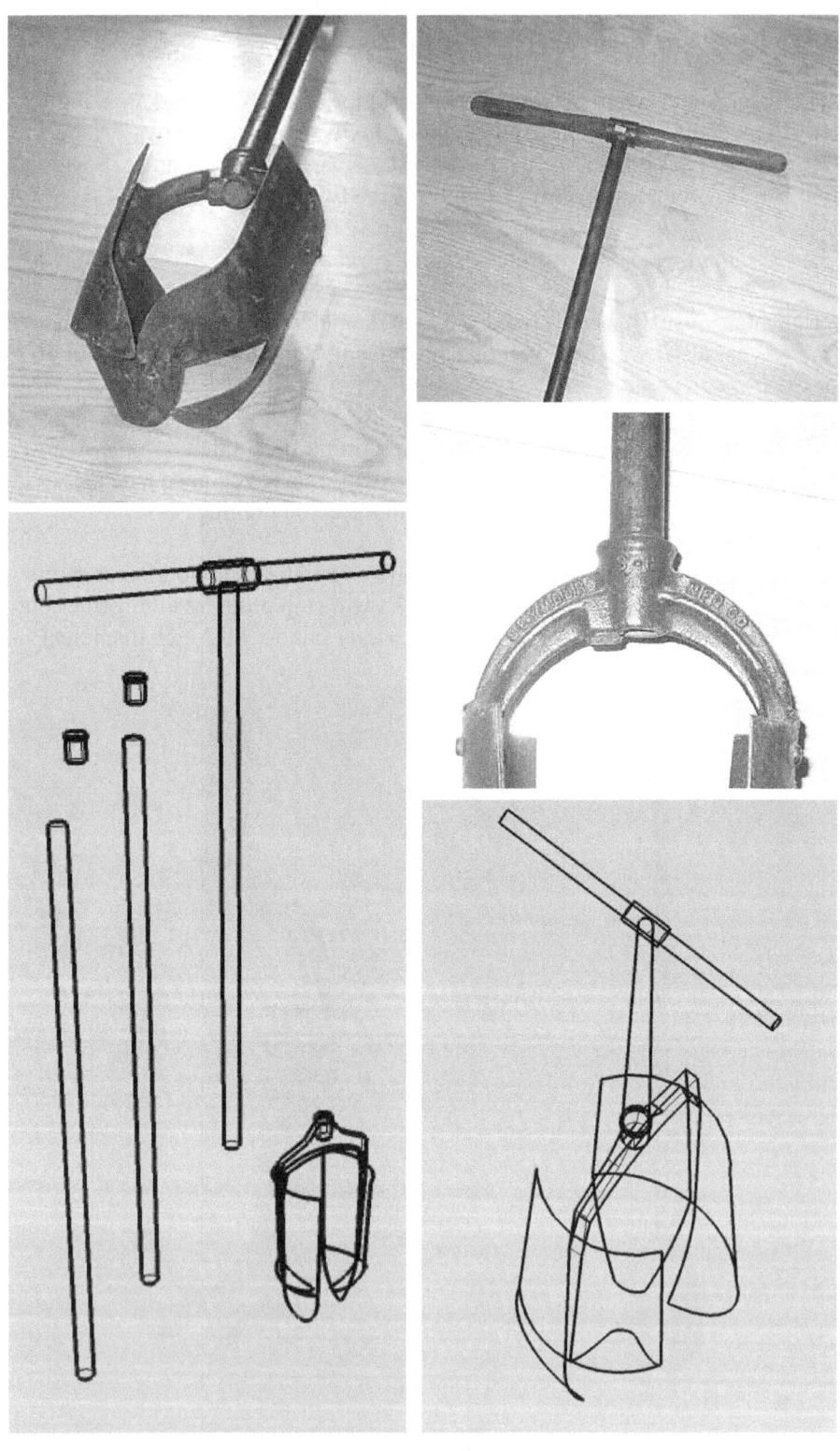

Hand Auger Boring Procedure

To start boring with the hand auger you force the auger blades into the soil while turning the tool. The auger will cut into the ground at a rate determined by the hardness of the soil. When the space between the blades is full of material, remove the auger and empty it. Repeat the operation until you reach the desired depth.

Connecting several lengths of pipe results in a rather heavy load to lift (each time the auger head is raised for emptying). Using two pipe wrenches (set to loosely fit the pipe) makes it easier to lift a heavy auger assembly from the hole.

Additional leverage can be gained by using two large pipe wrenches (gripping the auger shaft) with pipes placed over the wrench handles creating levers. The longer wrench handles created by the pipes will generate an incredible amount of leverage, so much so that it is possible to break the auger shaft or pipe. Use of this technique to turn an auger head that is seized on a rock will result in certain breakage. Pipe wrenches and leverage pipes may be useful for turning augers with long extension pipes that require more torque than the standard handle is able to generate.

If you encounter a large rock while augering, or the auger head becomes stuck or breaks off in the hole, you may have to abandon the hole and start somewhere else. The standard procedure is to simply withdrawal tooling and fill in the abandoned hole, but if you have a percussion drilling setup you can remove the auger and use a percussion bit to break up troublesome rock.

Traditional spiral augers are great for cutting a bore hole, but are not very efficient at containing and lifting material out of the bore hole. The Iwan auger was a big improvement in auger design because it contained cut material better within is bucket design for lifting to the surface. With this in mind you may want to add a bailer and the use of drilling mud (water and bentonite) to your traditional auger procedure.

Left: when extending the Iwan auger with standard 3/4 inch pipe, make sure to utilize a solid coupler as seen at the far left. The solid style coupler will withstand much more torque than the thin cast coupler which can crack under stress.

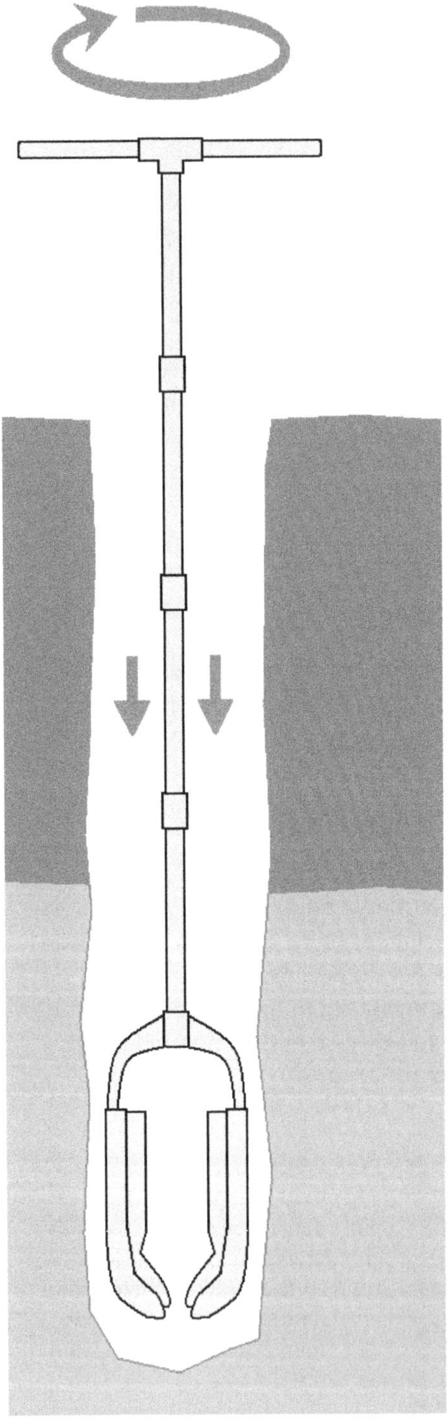

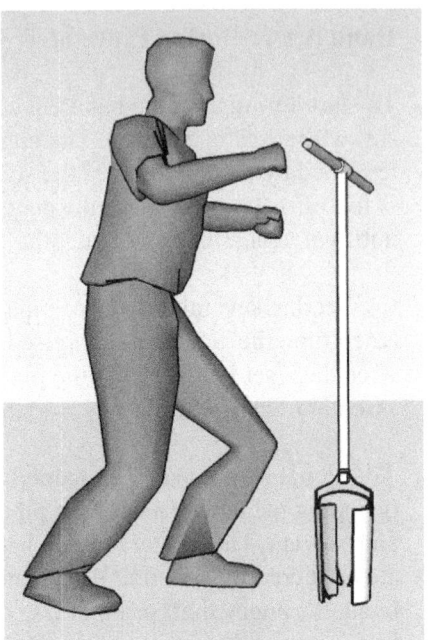
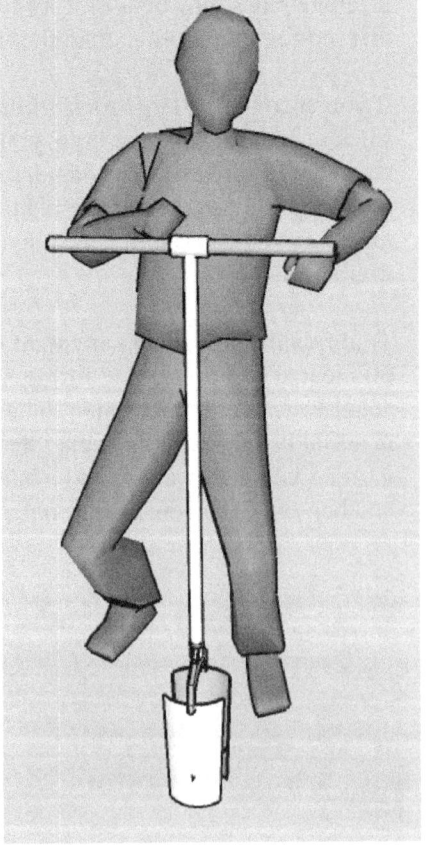

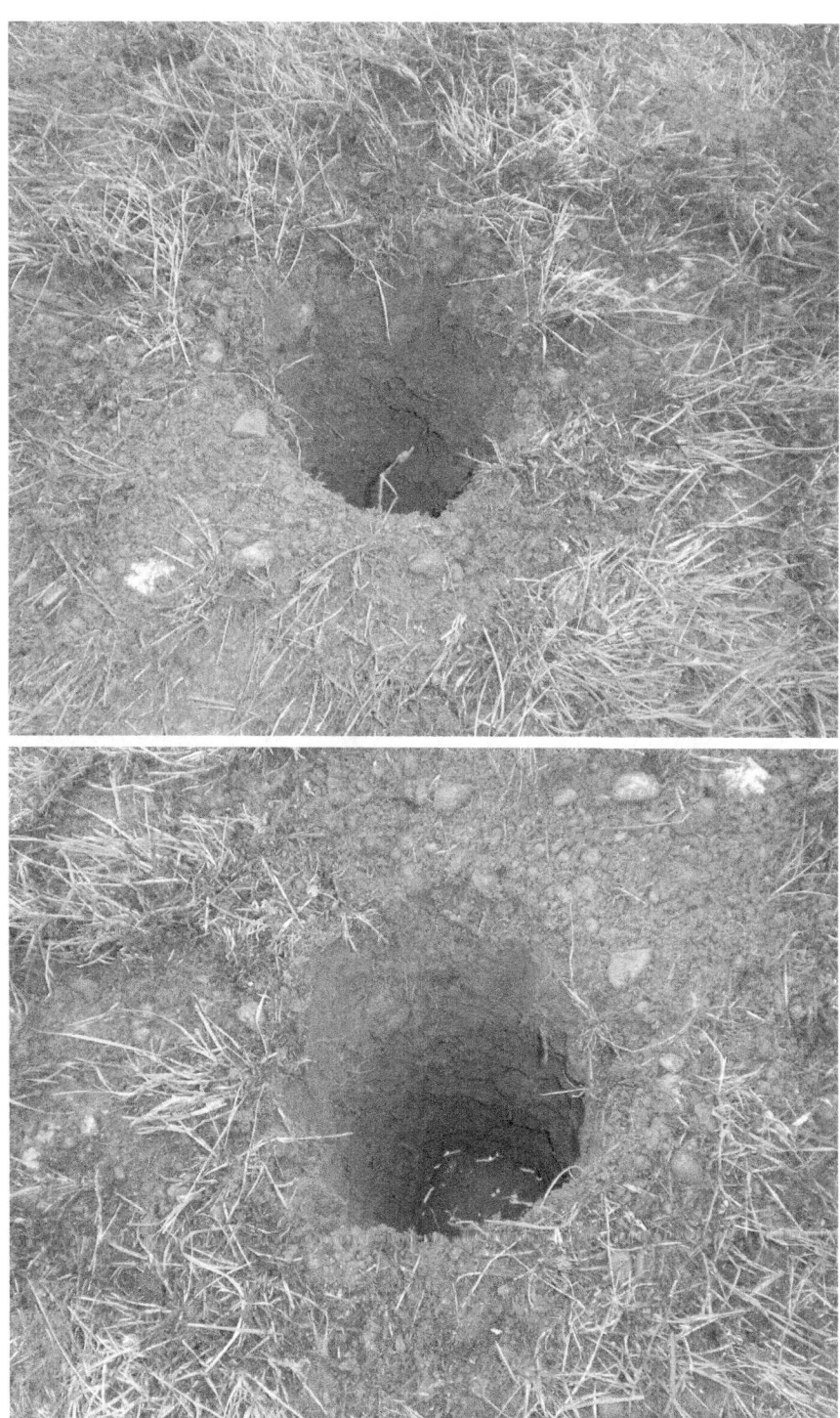

Above: bore hole created using the Iwan Auger awaits casing install.

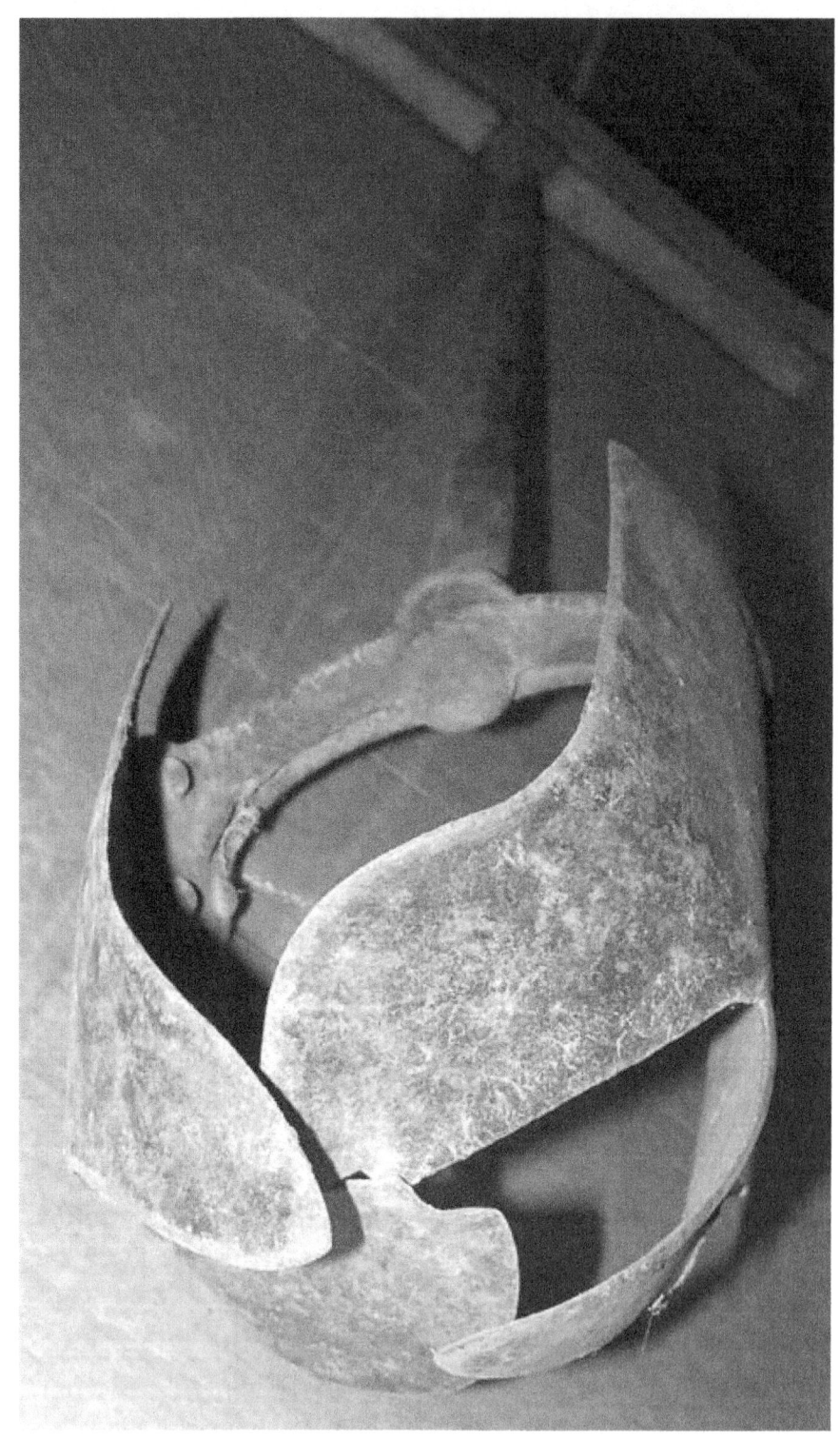

Iwan Auger

Drive Points for Creating Ready Made Wells

Overview

Low tech and low cost wells can be created in high water table areas with soft ground conditions using what is commonly known as a 'drive point', 'well point' or 'sand point'. The drive point, pipes and materials are purchased ahead of time and are usually not fabricated by the driller, although, in years past fabrication was common. The drive point method was often used by farmers and ranchers for creating wells on isolated areas of acreage for irrigation and livestock.

Drive point wells are installed by pounding the point and pipe assembly into the ground. The ground should be free from rock, hard clay, and other solid formations for successful drive point installation.

Drive points feature a cast iron point attached to a pipe embedded with a stainless steel mesh jacket and gauze screen. They are well drilling tools, casings, and well screens all together in one pre-made unit.

The two accessories sold alongside drive points are drive point couplers, and drive point caps. These are specially designed from extra heavy duty steel permitting the body and threads to withstand the pounding involved in drive point installation. The drive point cap has a hole for inserting wire or string to periodically check for water, and to allow free flow of air during driving. Both caps and couplers have standard pipe fittings threads for the stated drive point size.

Some drive points come with one coupler already attached, and some are threaded to accept a coupler but do not include one. Well points are generally sold in two diameters; 1-1/4 inch and 2 inch, and two lengths; 24 inch and 36 inch.

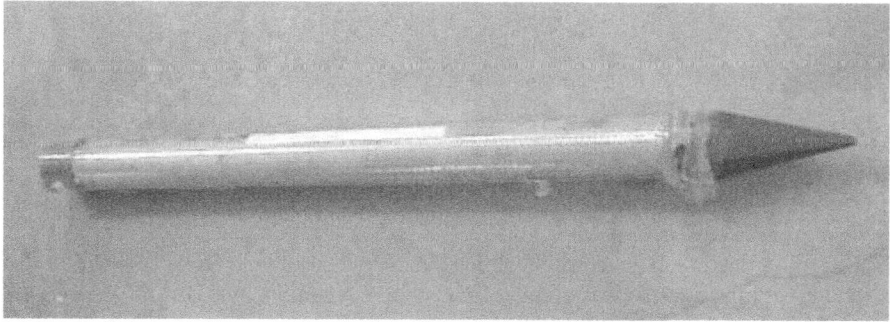

Above: new drive point assembly

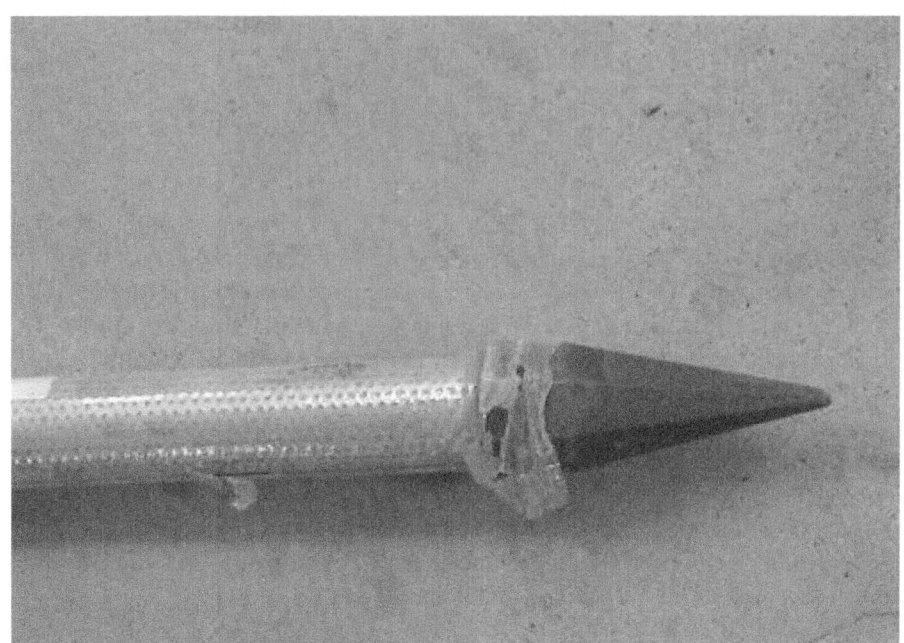

Top: front end of drive point is made from a solid cast iron piece of steel designed for maximum penetration while being driven. Bottom: back end of drive point featuring a welded-on coupler for connecting additional segments of standard 1-1/4 or 2 inch galvanized pipe.

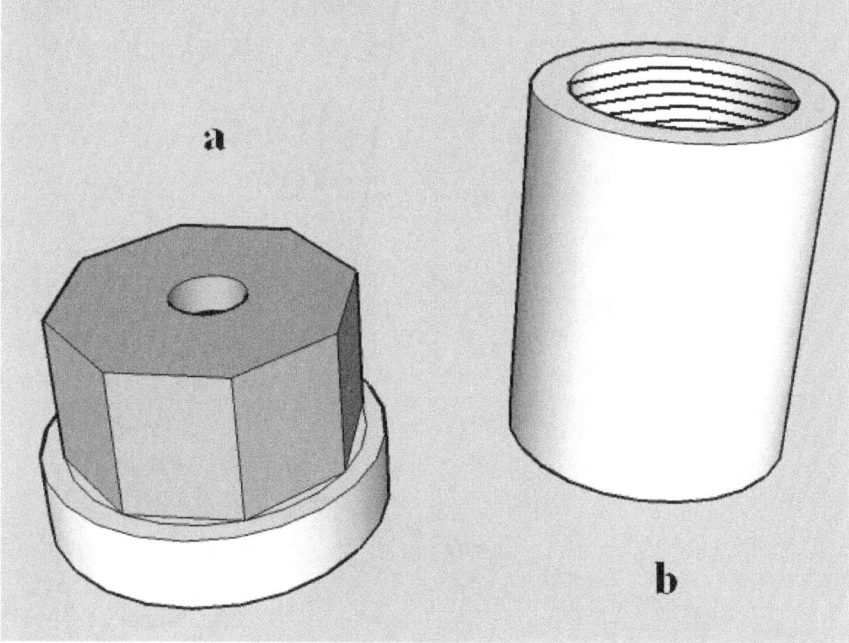

Top: regular style cast pipe coupler (left) shown next to a special solid drive point coupler (top right and bottom image b). The regular cast steel coupler can crack under the stress of repeated impacts involved with drive point installation.

Bottom a: special drive point caps are available for purchase along with the drive point. These are made from a thick impact resistant steel. Cord or wire can be inserted into the pipe through the cap hole to check for the presence of water during the installation process. Drive caps and couplers feature standard pipe fitting threads. Teflon tape should be used, and all joints tightened to maximum levels for prevention of air leaks that will cause pumping failure on these type of wells.

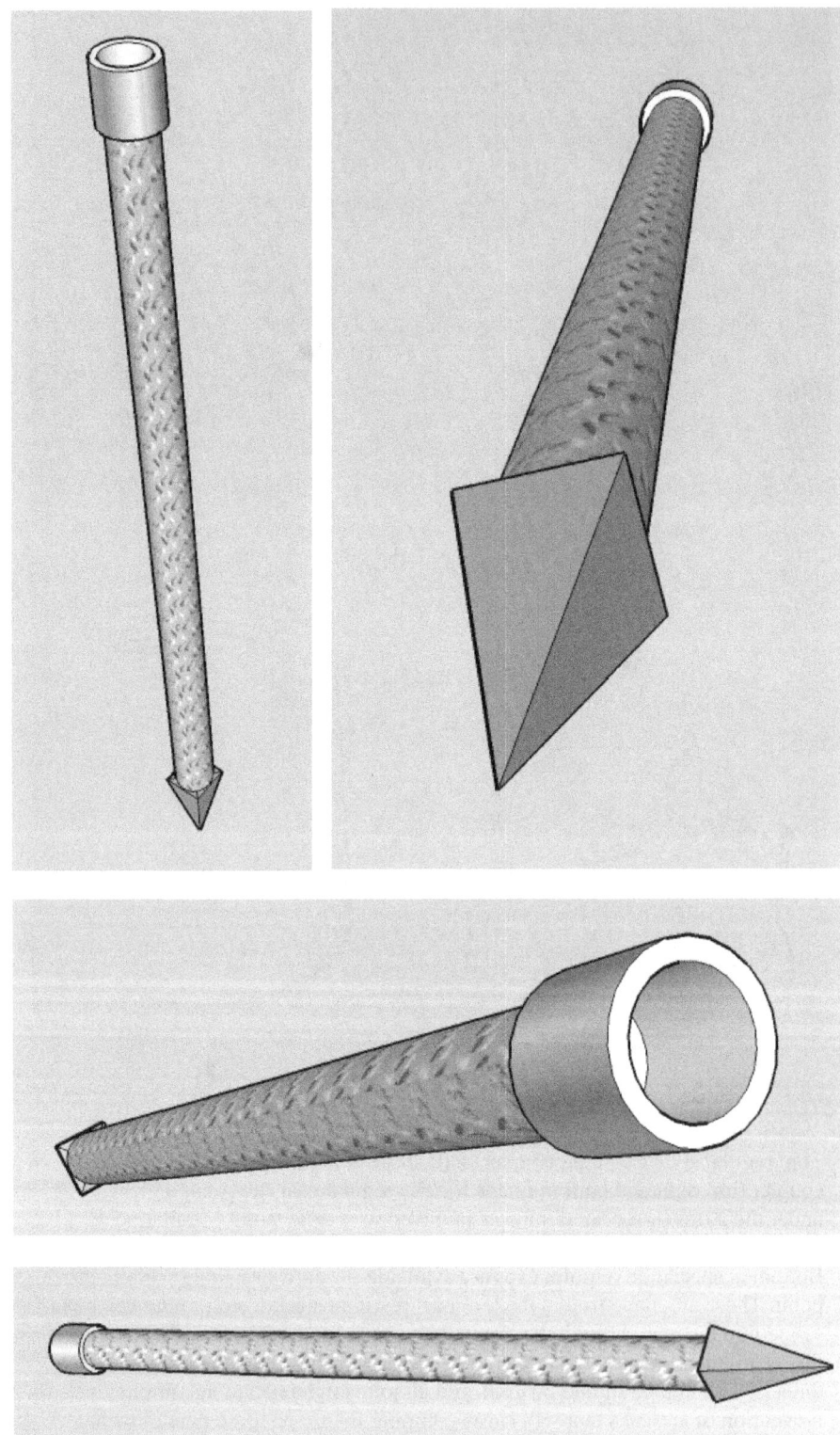

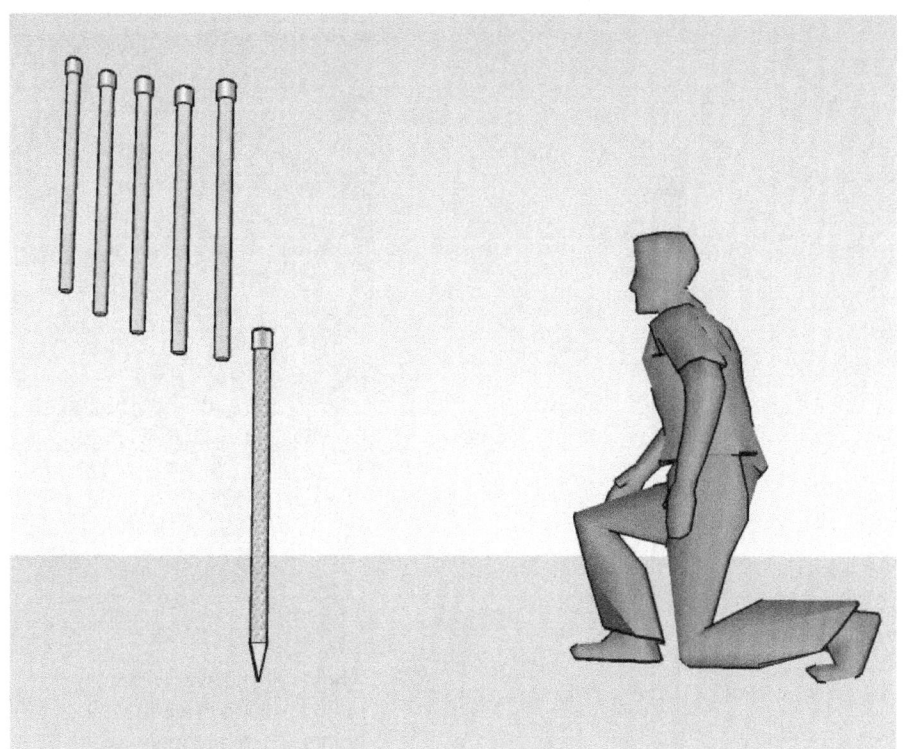

The standard method for installation is to drive down 'three pipes and a point', meaning three 5 foot galvanized pipe segments with the drive point at the tip. The drive point itself it usually 3 feet in length, resulting in a shallow driven well 17 feet deep.

Four or more joints can be used if the operator is reasonably certain such depths are attainable (keeping in mind that the pumping limit for shallow wells is 25 feet).

The pipes must be connected using the special drive point couplers. The impact of blows used to drive the point is absorbed with the special drive point cap installed on the end of driven pipe segment.

The installation of a driven well begins by boring a hole with a hand auger or posthole digger as far as possible. Connect a comfortable working length of pipe to the drive point and drop it into the hole. The drive cap is then placed on the top of the exposed pipe and a weighted driver is used to strike the drive cap, driving the point into the ground. When the drive cap is driven close enough to the ground and driving cannot be continued, another length of galvanized pipe is added and driving is resumed. Air leaks in the joined segments (from loose couplers) will cause pumping failure in this type of well, so use two pipe wrenches and torque each segment to maximum levels.

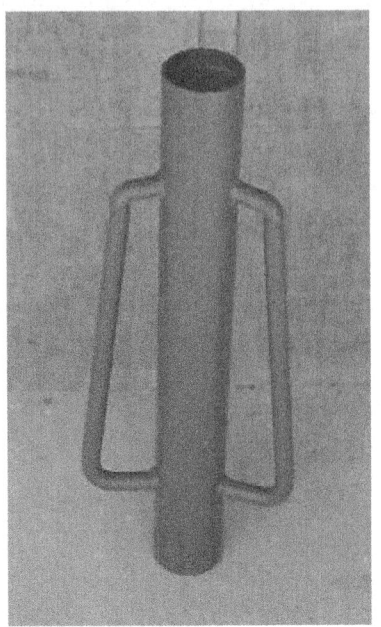

Above: post drivers are commonly used to drive in well points. There are many types available from rental facilities ranging from plain hand held manual drivers (as seen above), to air operated pneumatic drivers.

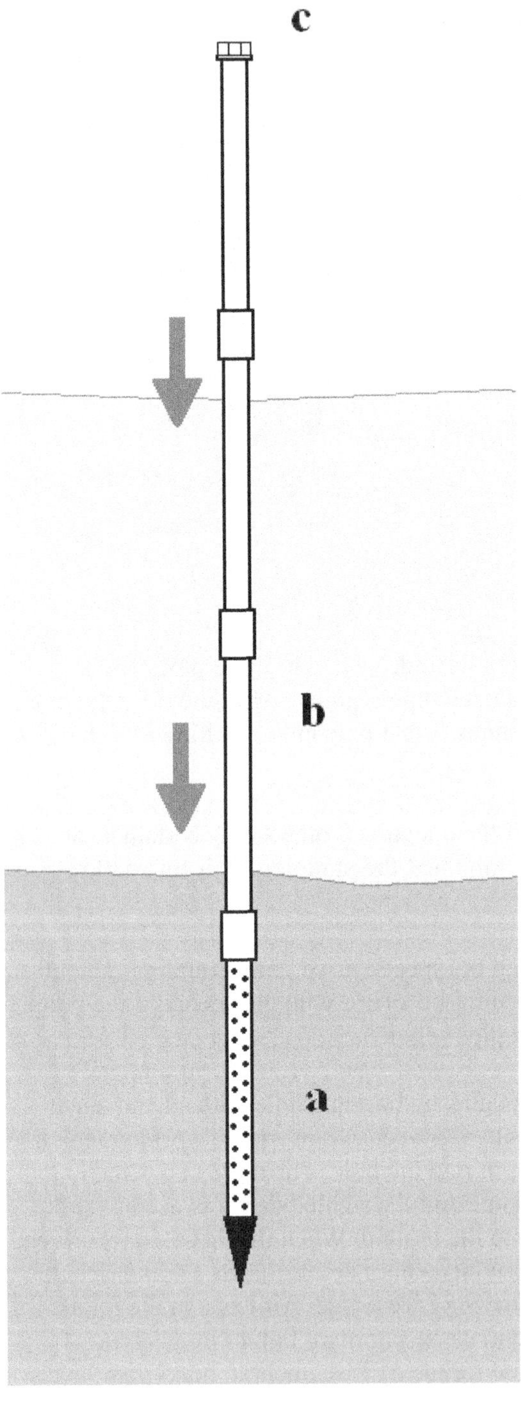

Figure a: drive point with it's built-in screen reaches water bearing ground formation.

Figure b: segments of regular galvanized pipe (usually 3-5 feet long) are connected via drive couplers creating the main structure of the well (the casing).

Figure c: exposed top of pipe with drive cap. Cap is removed when final depth is reached and a pump is attached.

Easier installation of drive points can be achieved by fabricating a special weighted driver made from a 2-3 foot long piece pipe with a 8-15 inch length of (round or square) solid steel welded to the top end for weight. Drive caps may vary in size so you will need to measure your drive caps outside diameter and use a steel pipe that will fit over the drive cap without too much play. Make sure to cut the pipe so that it stops short of contacting the couplers on each joined pipe. Weight of the driver will depend on the diameter and length of the pipe and weighted segment used with the minimum for hand held operation being 25 lbs.

This driver acts like a sheath fitting over the drive point pipe in a slide hammer fashion as the operator lifts and drops it by hand allowing the weighted end to do the driving work. Adjustments such as the addition of larger weighted segments may need to be made if inadequate penetration is observed. This type of driver can also be suspended from a tripod and strike the drive point assembly while being lowered on rope and pulley, useful if the driver weight exceeds 30-40 lbs.

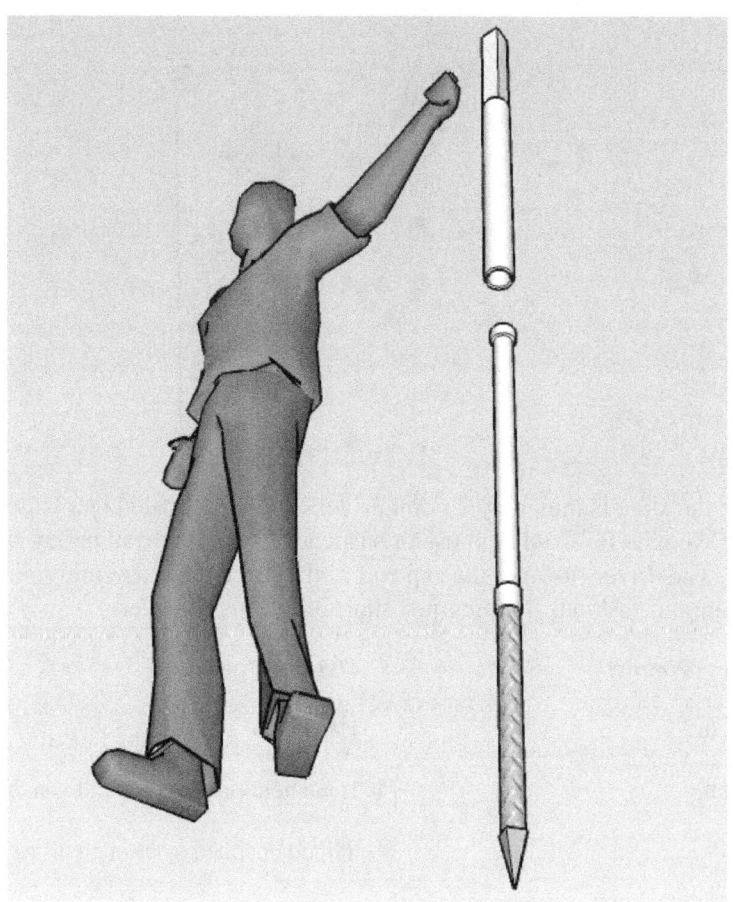

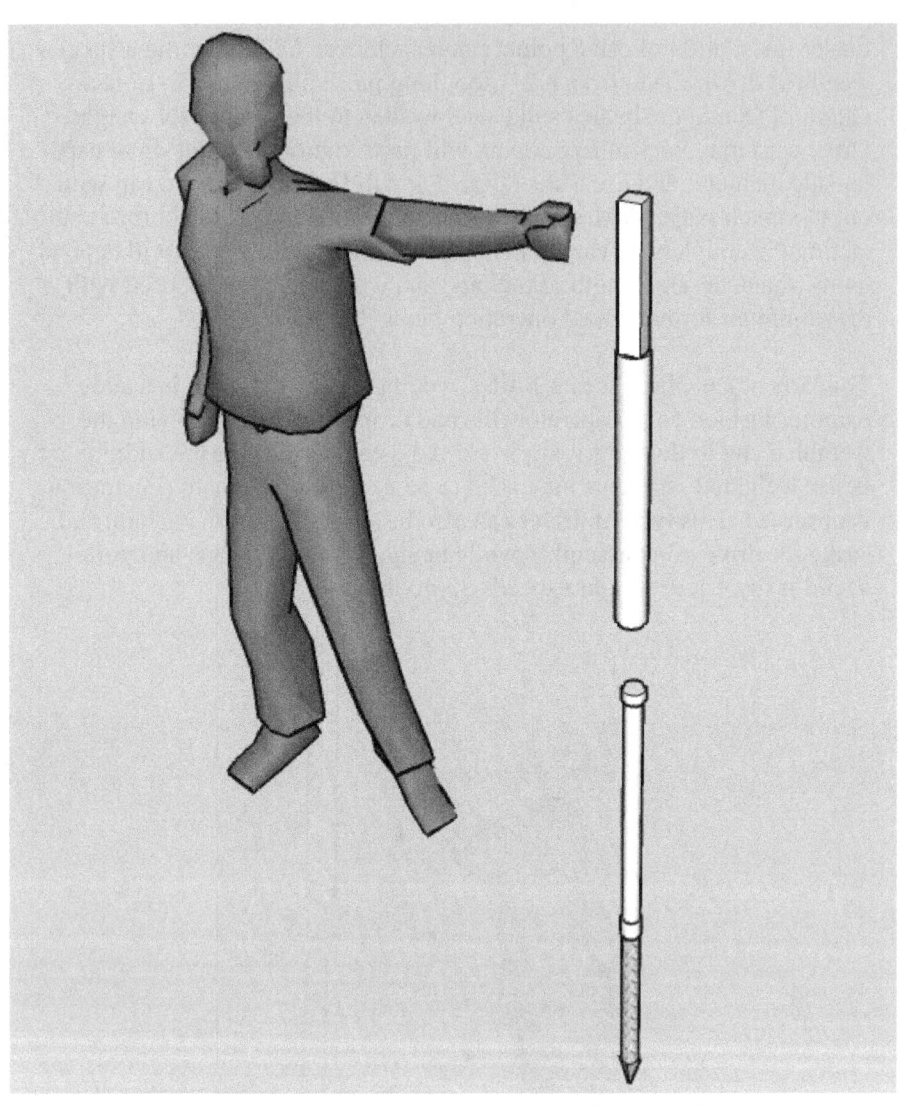

Above: the slide hammer style point driver can be fabricated in a wide variety of weights for hand held use or suspended by a rope and pulley via tripod. The driver fits over the cap end of the drive point assembly, in a slide hammer fashion, stopping just short of the first coupler.

Driver Weight	Driver Operation
10-20 lbs	Hand held.
20-30 lbs	Hand held (install T– handle on driver).
30 lbs+	Tripod operated (rope and pulley).

Slide Hammer Style Point Driver

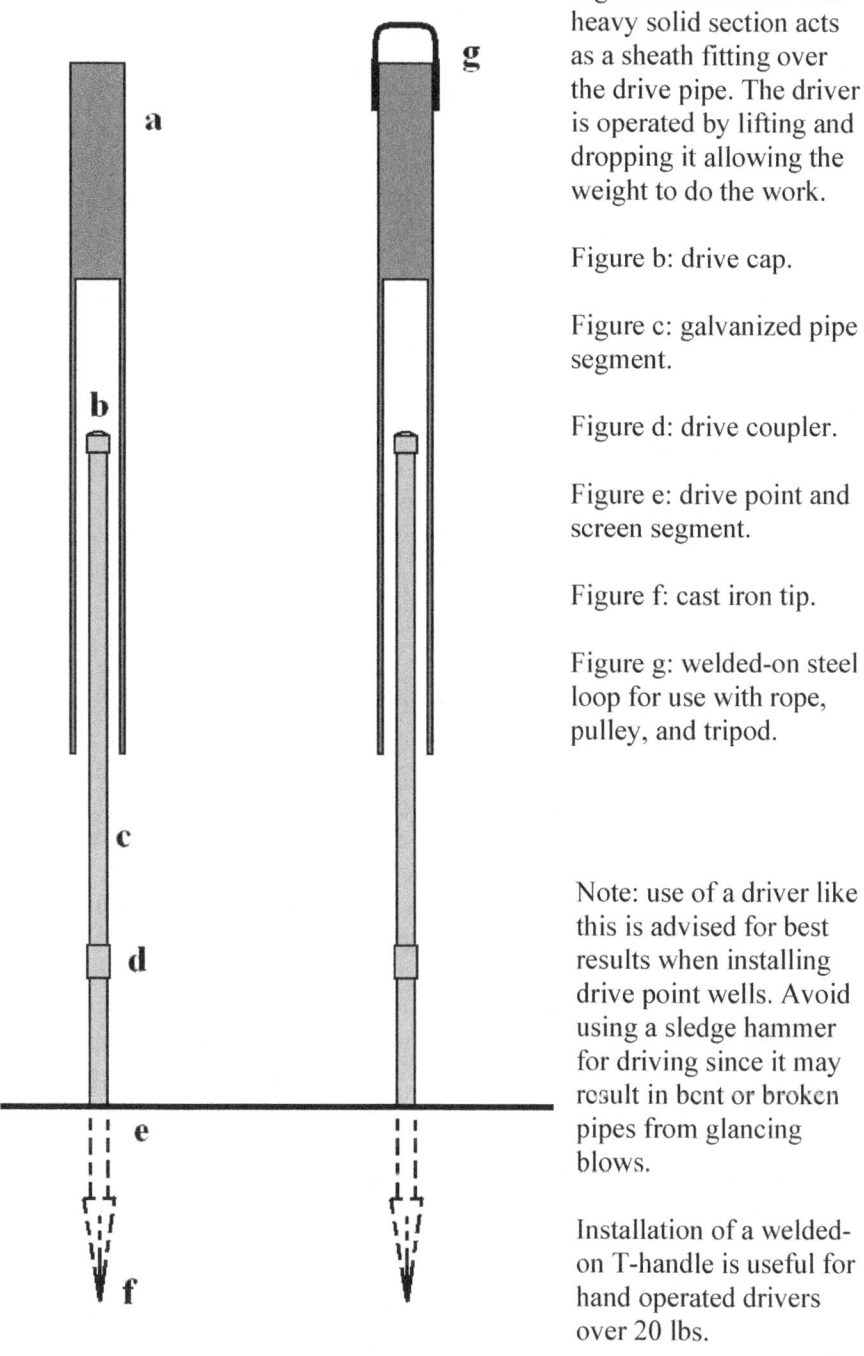

Figure a: steel tube with heavy solid section acts as a sheath fitting over the drive pipe. The driver is operated by lifting and dropping it allowing the weight to do the work.

Figure b: drive cap.

Figure c: galvanized pipe segment.

Figure d: drive coupler.

Figure e: drive point and screen segment.

Figure f: cast iron tip.

Figure g: welded-on steel loop for use with rope, pulley, and tripod.

Note: use of a driver like this is advised for best results when installing drive point wells. Avoid using a sledge hammer for driving since it may result in bent or broken pipes from glancing blows.

Installation of a welded-on T-handle is useful for hand operated drivers over 20 lbs.

Drive Point Installation Tips

° When adding each segment of pipe to the assembly, wrap about 3 flat layers of Teflon tape clockwise on the pipe threads of each drive coupler and tighten as much as possible. Use a second pipe wrench to hold the pipe when tightening the couplings. This is important because air leaks in loose coupler joints will cause pump failure with this type of well.

° When the water table is reached the drive point assembly (pipes) switch roles, and become the casing with the attachment of a pump for developing the flow of the well. Hand pumps must be mounted directly over the pipe.

° Drive points are sometimes used with augers in loose ground formations. If the auger cannot penetrate further into a loose sandy layer, the well point can be driven into the bottom of the existing (unfinished) bore hole achieving greater depth and providing a ready made screen and casing.

° Drive points can also be installed in the bottom of pre-existing wells that have gone dry often reaching water a few feet below the bottom of the bore hole. If the existing well casing and screen are PVC, it is possible to punch through the casing end cap with the drive point, otherwise the casing must be removed prior to drive point installation.

° Atmospheric pressure (of about 15 pounds per square inch) dictates that a shallow well pump can only lift water from a maximum depth of 25 feet.

° If the drive point reaches a level deeper than 25 feet, the 1-1/4 inch style (drive point) will fail to pump (assuming you are using a regular shallow well pump). It is recommended you use a 2 inch drive point if you believe such depths will be reached, but this is a 'catch 22' so to speak. The 2 inch drive point required for greater depths (for pumping reasons) is actually much more difficult to install to such depths. Ground resistance and friction makes installing any drive point past 25 feet difficult, and the 2 inch with it's increased diameter, requires more effort than the smaller diameter model. The 2 inch point past 25 feet is necessary because the drive point pipe must be large enough to accommodate a deep-well type jet pump able to pull water up past 25 feet. This may be a moot point since it's rare that a drive point can even be installed at depths greater than 25 feet. These possible post-installation pumping problems are further examples of why you should have an idea of ground water level before choosing your well creation method.

° Avoid using low quality pipe could bend or crack during installation. The pipe used should be steel galvanized and meet the appropriate ASTM or API standards: (ASTM A53 A106 A589) and (API 5L 5LX 5D 5CT).

Well Installation; Casing & Finishing Bored Well Holes

Overview

Until you install a casing and screen (the screen is affixed to the bottom end of the casing) into your bored hole, you really have only created a long narrow horizontal tunnel in the ground. This tunnel needs walls and a sediment screening section near the bottom, or in other words, a casing and screen.

The casing and screen make up the major material aspects of your well. Think of it like building a house that will have a basement. Excavation for the basement must be completed prior to pouring the concrete walls and floor. With well installation the boring of the hole and casing insertion is much the same. In well building, casing installation and casing insertion can be used as interchangeable terms because they really are one in the same.

Modern casings are almost always made out of common everyday PVC tubing available at your local home improvement store. While special screen sections are available for connecting to PVC casing, the low tech well will usually use the same size PVC pipe (used for the casing), except that it will have slots cut into it and a sediment filter sock placed over it.

Signs of Ground Water in Bore Hole

Boring a hole in the ground is one thing. Installing the well (casing, screen, and pump) is another, but what comes between these two tasks is reaching the actual ground water. This is the entire point after all.

Always check for water during the well boring process (whatever method you choose) at the depth your research has suggested. It's common practice to consult state water depth records (usually available online) for local water depths, or inquire with neighbors about well depths. When you encounter ground water you may notice a difference in the cuttings that might indicate water, or colder water in the bore hole, but the best indicator is to leave the bore hole overnight and check in the morning to see if there is any difference in water level.

Note: always cover bore holes with 4x8 weighted plywood when left unattended.

Drop in a 100 foot tape measure with something heavy, like a pipe wrench, tied onto the end. When it hits bottom the tape will go slack and you can measure against the side of the case to check your depth. Measure the water level in the evening, then again the next morning and compare the two. You must pump out the bore hole (if it is full of drilling fluid) before conducting water level measurements. If the bore hole is filling up with water you have reached some level of groundwater formation.

Well recovery rate measurement that will show how rapidly the well is recharged after consuming water (below):

° Measure to the top of the water from the top of the well casing.
° Run pump for 10 minutes.
° Re-measure to the top of the water.
° Wait 30 minutes and re-measure.

If your water level is close to where it was prior to running the pump then you have a good recovery.

Note: always position or install pump intakes at least 5-10 feet above the bottom of the well casing to avoid stirring up silt and sand that will damage the pump.

Casing, Screens, and Pumps

It's important to figure out how much water yield will be required for your well. This will decide how long of a screen section you will need. The longer the screen section (inserted into water bearing formation), the more water your well will produce. For example, a 20 foot length of 2 inch schedule 40 screen will hold 3-1/2 gallons of water. Your pump will remove and transfer this water rapidly to your pressure tank, bringing up the next point of concern which is how long it will take for the casing to be recharged (refilled with water). When you create a large filter pack (gravel between the bore hole and surrounding the screen) it acts as a water holder, keeping water at the ready to flow through the screen and refill your well quickly.

Schedule 40 PVC Casing Volume Capacity (thinner wall pipe holds slightly more)

2 inch PVC holds .17 gallons of water volume per linear foot.
4 inch PVC holds .65 gallons of water volume per linear foot.

Schedule 80 PVC Casing Volume Capacity (thicker wall pipe holds slightly less)

2 inch PVC holds .15 gallons of water volume per linear foot.
4 inch PVC holds .59 gallons of water volume per linear foot.

The 4 inch well (4 inch PVC casing tube) is the most common in North America. These wells are most often equipped with a submersible pump that is lowered into the well and secured with a safety line. They are the industry standard; quiet, efficient, never loses prime, but if pump servicing is required, extracting the pump can be time consuming. The 4 inch pump is actually 3.5 inch in diameter to fit inside the casing. Also worthy of note is the high price of submersible pumps.

The 2 inch well (2 inch PVC casing tube) is the standard for shallow wells and is called for in many do-it-yourself well projects. Tubing cost less but the obvious trade off is less water volume delivered from finished wells. It is also easier and faster to bore small diameter holes. Wells made with 2 inch casing can be pumped from above with a jet pump. Shallow (under 25 feet) and deep well jet pumps (under 100 feet) draw water up from the well. These type of pumps make a lot of noise during the pump cycle, use more electricity, and can lose prime on the foot valve down in the casing. Pump and all equipment is above ground for easy repair access. Old fashioned hand pitcher pumps or 12 volt solar pumps can also be fitted to shallow wells for off-the-grid use.

Well casing and PVC are almost the same product, save for casing branded material having a higher heat rating to accommodate heat generated so close to the tube wall by submersible pumps. Any well that utilizes above ground jet pumps can safely use regular PVC for casings. Standard PVC pipe is available everywhere, but casing branded PVC is only available from plumbing supply dealers. Most PVC pipes are available with bell-housing couplers built in. Professional well drillers like to use 20 foot lengths pieced together to attain full casing depth, but you should use those lengths of PVC that you can easily transport (10, 12 foot etc).

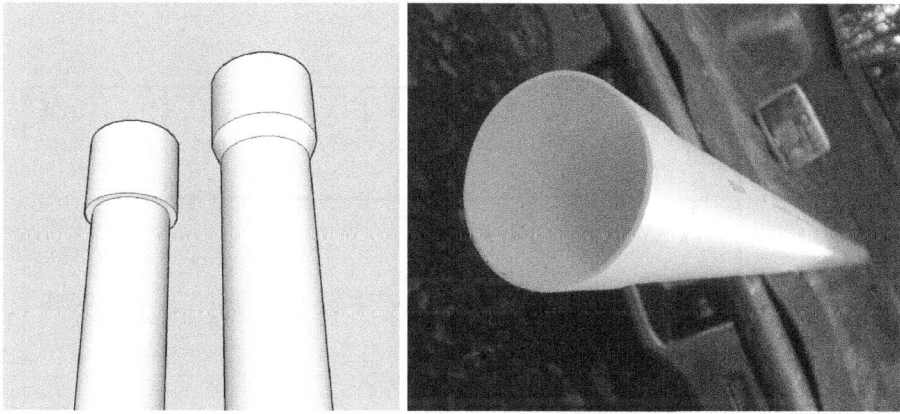

Left: PVC pipe with coupler and PVC pipe with built-in bell housing.
Right: bell housing end of 3 inch thin wall PVC plumbing pipe.

Schedule 40 and 80 PVC Pipe

The chart below shows the dimensions of the two most common types of PVC pipe; schedule 40, and schedule 80. The difference between the two is wall thickness, a factor that translates into a higher or lower maximum PSI working pressure. Schedule 80 PVC pipe is stronger with a thicker wall. Schedule 80 is heavier and more expensive than Schedule 40. Higher schedule number equals a thicker wall pipe, lower schedule number equals a thinner wall pipe.

PVC pipe is manufactured in strict compliance to ASTM D1785 and D2665 to meet quality assurance test requirements with regard to material, workmanship, burst pressure, flattening, extrusion quality, and so forth.

Schedule 40 PVC

Pipe Size Diameter	Outside Diameter (OD)	Inside Diameter (ID)	Wall Thickness	Weight Per Foot	Low Tech Application
1-1/4 inch	1.660	1.360	0.140	0.450	Minimum well casing
2 inch	2.375	2.047	0.154	0.720	Small well casing
3 inch	3.500	3.042	0.216	1.488	Medium well casing
4 inch	4.500	3.998	0.237	2.118	Large well casing

Schedule 80 PVC

Pipe Size Diameter	Outside Diameter (OD)	Inside Diameter (ID)	Wall Thickness	Weight Per Foot	Low Tech Application
1-1/4 inch	1.660	1.255	0.191	0.586	Minimum well casing
2 inch	2.375	1.913	0.218	0.984	Small well casing
3 inch	3.500	2.864	0.300	2.010	Medium well casing
4 inch	4.500	3.786	0.337	2.938	Large well casing

Thin Wall PVC Drain Pipe Low Cost Casing Option

If you are working on a budget or just don't want to pay for premium schedule 40 or 80 PVC pipe, sewer and drain pipe is an option. These are thin wall PVC pipes used mainly for landscaping drainage. The brand I use is called 'Genova PVC Plumbing & Sewer Pipe 3000 lb Crush Resistant'.

This type of pipe makes a lightweight easy to handle and low cost well casing. PVC plumbing and sewer pipe is rated for crush resistance from external force as opposed to PSI resistance from internal (water pressure) force.

The thin wall is also much easier to cut slots into when fabricating casing screen segments. Thin wall pipe should not be used with a planned submersible pump setup due to the heat these pumps generate so close to casing pipe wall.

PVC thin wall sewer and drain pipe has built in bell housings, is about a 1/3 of the cost of thicker PVC pipe, and offers a workable substitution for schedule 40 & 80 PVC. This product is sold in ten foot lengths of 3, 4, and 6 inch diameters. Make sure to use the correct matching drain cap for this style pipe at the bottom of the screen segment. Drain and sewer pipe fittings are somewhat smaller than PVC supply and ABS DWV pipe fittings in diameter and require special caps and fittings.

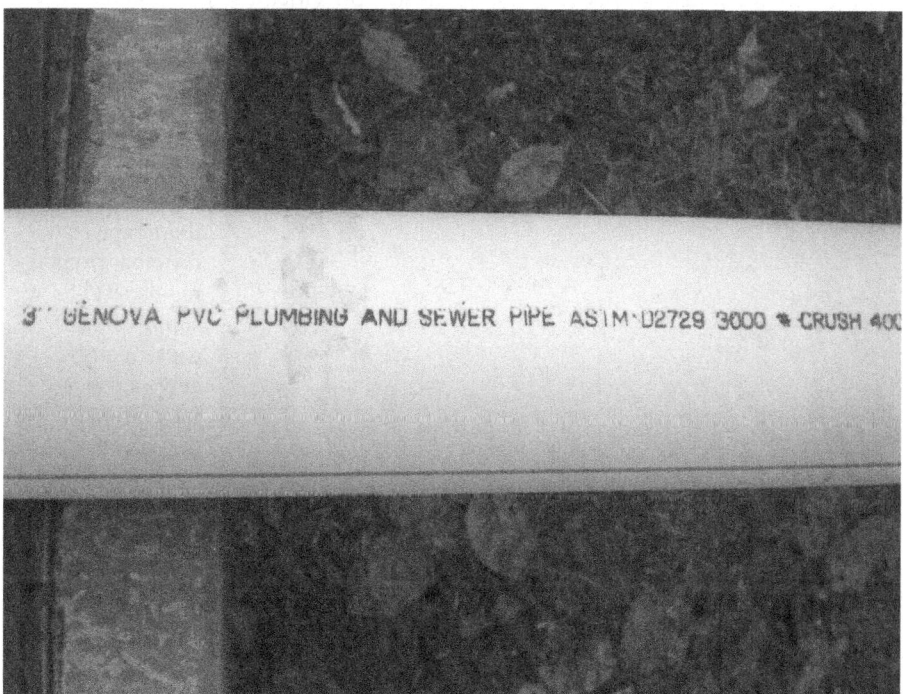

Solid PVC Sewer and Drain Pipe (Thin Wall)

Pipe Size Diameter	Outside Diameter (OD)	Minimum Wall Thickness	Weight Per Foot
3 inch	3.250	0.069	0.478
4 inch	4.215	0.073	0.660
6 inch	6.275	0.093	1.266

Note: Black ABS plastic pipe designated 'DWV' (Drain, Waste & Vent) is a different product than Genova white PVC Plumbing & Sewer Pipe.

Note: Black ABS plastic pipe designated 'DWV' (Drain, Waste & Vent) has the same outside dimensions as PVC supply pipe.

Note: PVC and ABS plastic pipes require the correct adhesives that are sold near pipe.

One final word on what kind of pipe to use; essentially any brand or style of modern PVC plumbing pipe will work as a low tech well casing. Gray PVC conduit pipe can even be used as casing and screen because it is tough and will accept standard supply and DWV fittings provided the proper adhesive is used.

Left: cap for 3 inch thin wall PVC sewer and drain pipe is different from regular PVC supply pipe caps. Wells receive a cap at the bottom of the screen segment which lies at the bottom of the bore hole. Glue it on using the correct PVC adhesive.

Well Screen Fabrication and Preparation

The well screen is the lower segment of casing that features many slots or holes to allow water in and keep sediment out. Ten to fifty percent of a well depth is usually comprised of screen, but this depends on water depth. Take for example a 100 foot borehole with a PVC casing inserted into it. The water level in the hole may only reach up 20 feet from the bottom, so you only need a 10- 20 foot screen segment (at the bottom) and the remaining 80 feet can be solid PVC tube all the way up to the surface. If you have a 100 foot well hole with a higher water level in it, say 40 feet, you will lose out on available water volume by not installing a longer (20-40 foot) screen section.

Well screen products (standardized for PVC connection: stainless steel mesh, pre-made gravel pack, sand pack well screens, etc) are available for purchase from well and plumbing supply houses, but these items can be costly.

Low tech homemade well screens are much cheaper and almost as effective. They can be easily fabricated from standard PVC pipe using a dremel tool or hacksaw. Lay a 10-20 foot piece of PVC pipe on a workbench or truck tailgate and cut 12-15 rows of 2 inch vertical slits in it, leaving 6 inches on each end uncut. Make enough slits or holes to allow sufficient water into the pipe but not so many as to weaken it's structural integrity. You will also need a PVC pipe cap for the bottom end. If a slotting tool is not available, you can drill the pipe full of 1/8 inch holes.

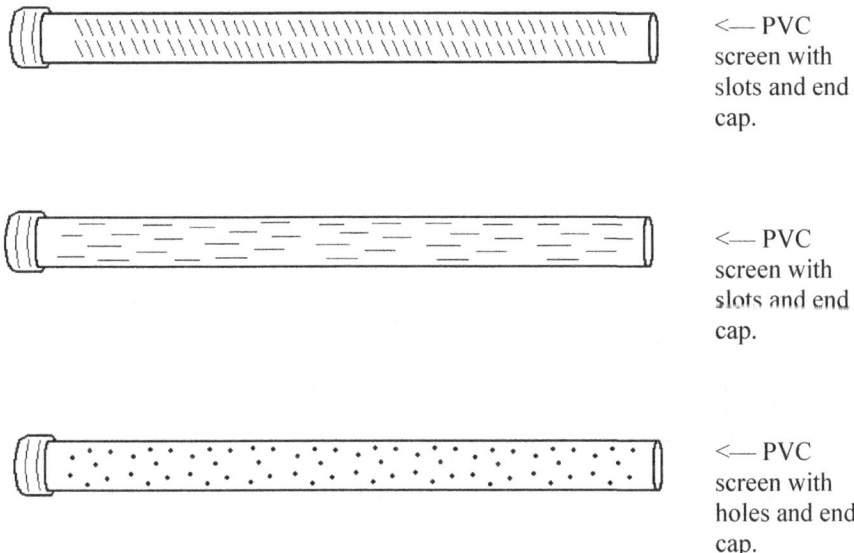

<— PVC screen with slots and end cap.

<— PVC screen with slots and end cap.

<— PVC screen with holes and end cap.

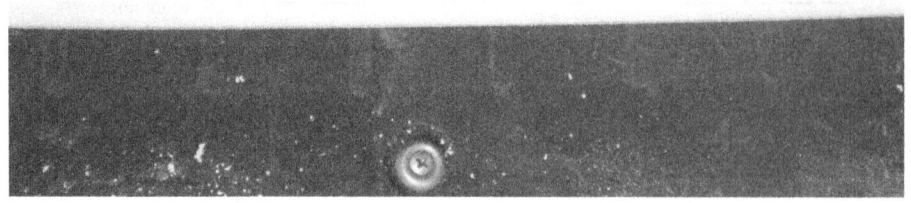

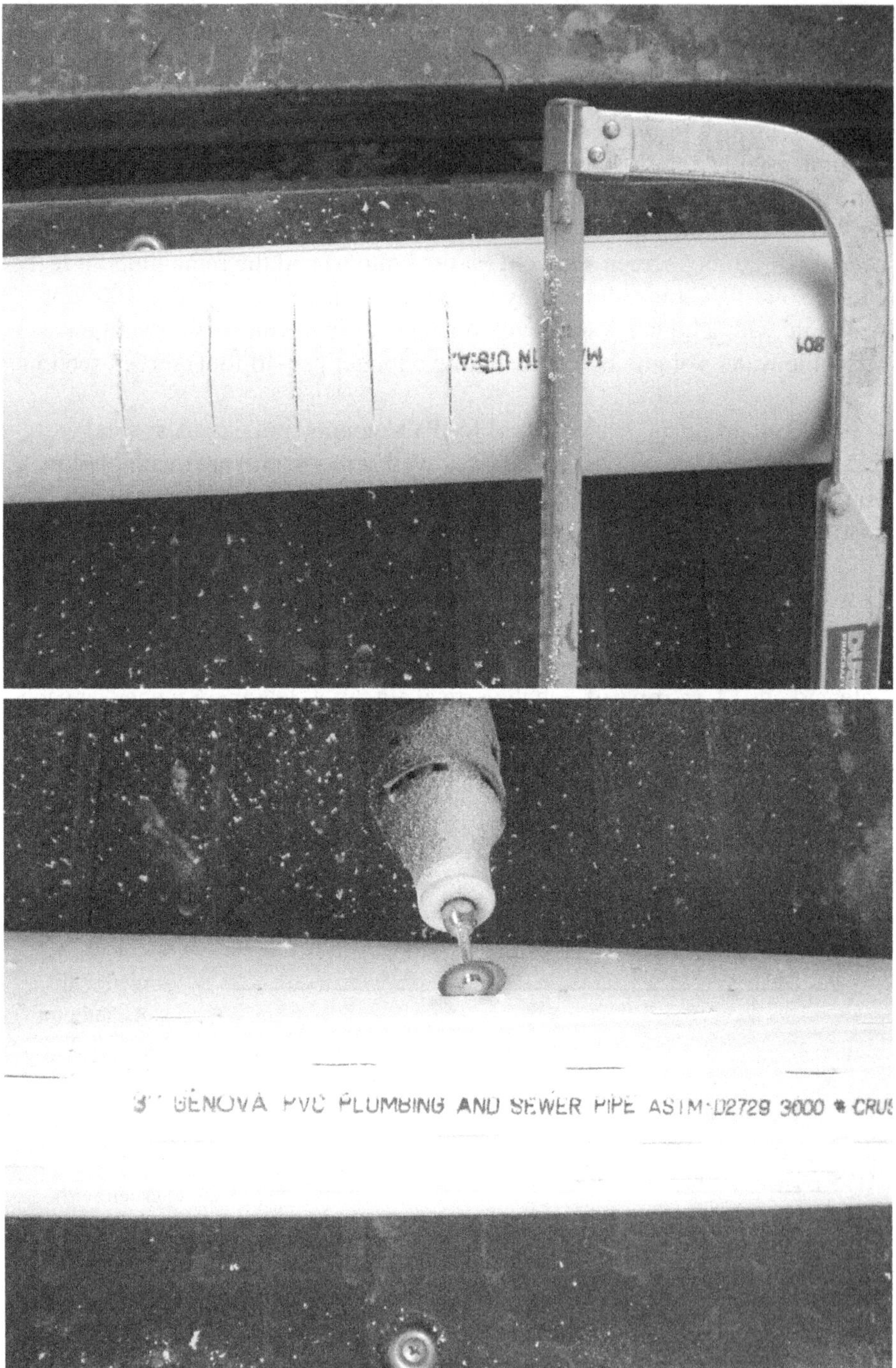

Top: hacksaw blade efficiently cuts slots in PVC pipe for screen.

Bottom: dremel tool with cut-off bit rapidly cuts slots in PVC pipe.

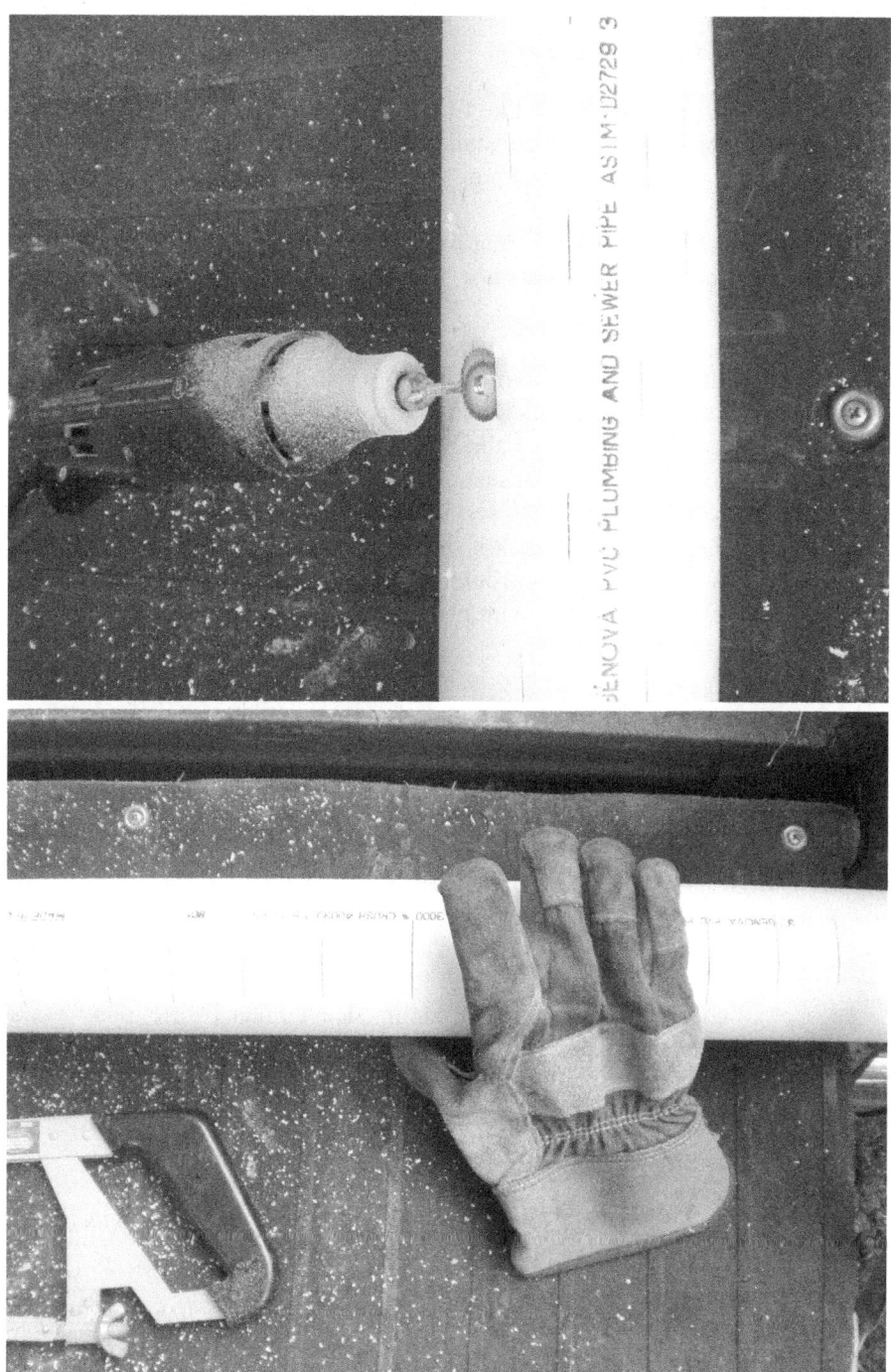

Cutting slots in PVC throws up burs that should be cleared away to prevent any blockages. Use a leather glove and rub the pipe to knock off burs. Ensuring that the pipe is smooth allows for snag-free filter sock application and maximum water flow.

Sediment Sock

The sediment sock is a sleeve-like filter fabric that is added to cover the slots on the screen pipe before installation in the bore hole. This keeps sand and sediment from entering the well.

Many different types of porous not-rotting fabrics can be used as improvised sediment socks, but they must be able to withstand some possible trauma during installation.

The first product that comes to mind is the screen sock material commonly sold for covering 3 and 4 inch flexible perforated landscape drainage pipe. 'Drain-Sleeve Filter Fabric' sock is available at most home improvement stores in the garden section. It's designed for filtering drainage pipe but can be pressed into service for an improvised low tech well screen. The 4 inch diameter sock is the only size I have seen for sale locally. It will work on 4 inch and smaller PVC screen pipe. You will need to wrap loose sleeve material with plastic zip ties for pipes smaller than 4 inches.

'Aqua-Sleeve Fabric' sock is another material you might use. It's a better product than regular drain sleeve because it's designed for well screens and comes in a variety of diameters other than 4 inches. It's a continuous seamless filter fabric knitted specifically for filtering ground water. Aqua-Sleeve filter fabric is available in two weights: regular knit (3-36 inch diameter) and plush fabric (1-12 inch diameter). It's resistant to scuffing and tearing during the installation and well development procedures.

The Carriff company produces both 'Drain-Sleeve Filter Fabric' and 'Aqua-Sleeve Fabric' packaged in assorted length socks. Aqua-Sleeve well screen fabric socks will be more difficult to find locally, requiring the possible placement of an online order.

Other similar products are also available from well drilling supply houses via the internet. Internet search keywords: "filter fabric sock", "drain sleeve filter fabric sock", and "filter fabric".

Do not use cheap perforated plastic landscape fabric as a screening material. It will degrade within 5 years. Woven landscape fabric is a better choice (if the filter sock is not available) and is commonly rated for 20 years. Fabric is usually sold in rolls 2-3 feet wide and can either be wrapped around the screen pipe or sewn into a sock. Internet search keywords: "woven landscape fabric".

Above: clean water inside your well casing is essential. 'Drain Sleeve Filter Fabric Sock' for perforated pipe prevents clogging and keeps water flowing. It's available at most home improvement stores.

Top: solid cap is attached to screen end using PVC solvent cement.

Bottom: filter sock is cut to size prior to installation on pipe.

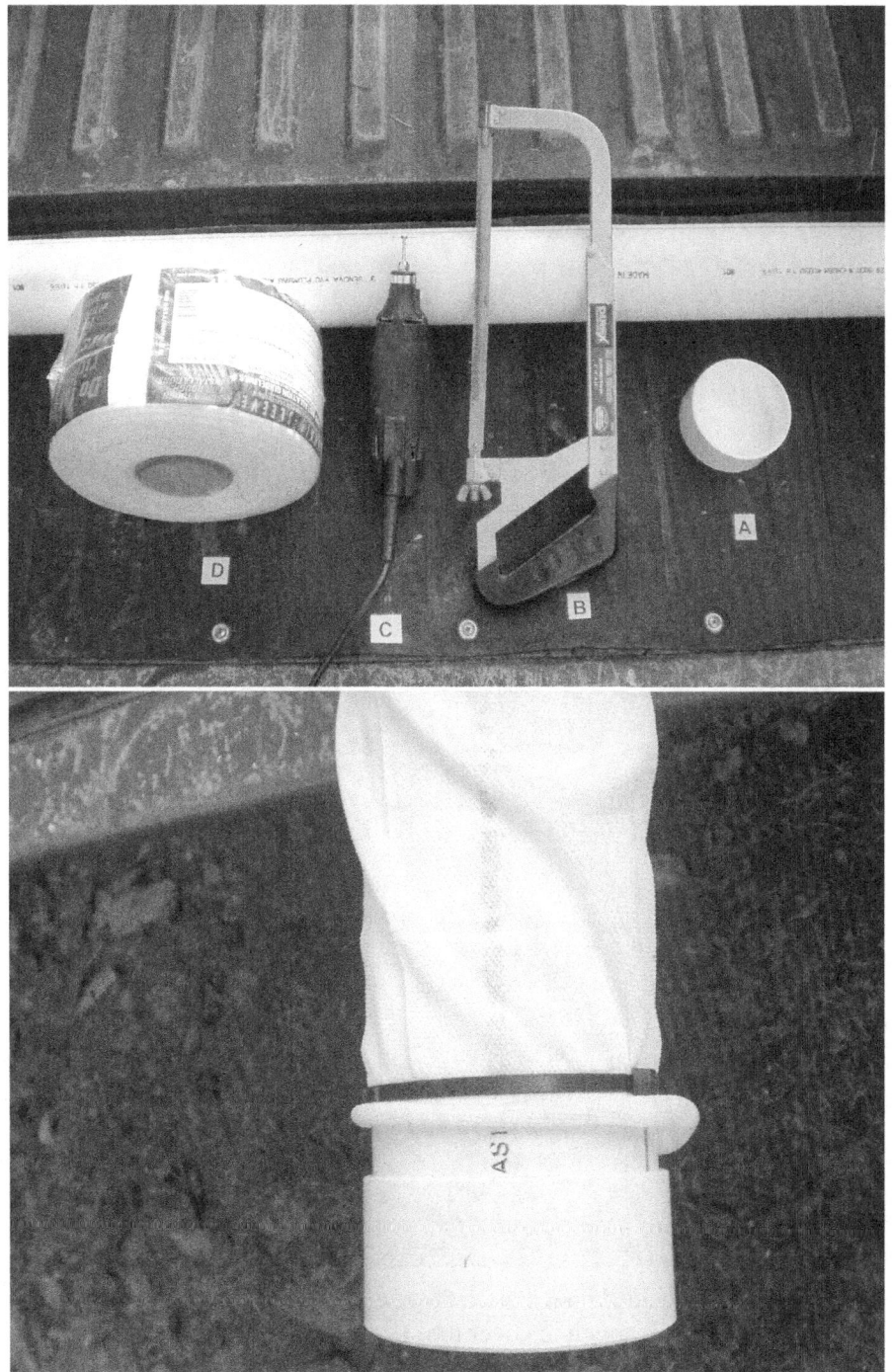

Top: screen fabrication parts and tools insure the job is done correctly.

Bottom: filter sock secured to PVC screen pipe with plastic zip ties.

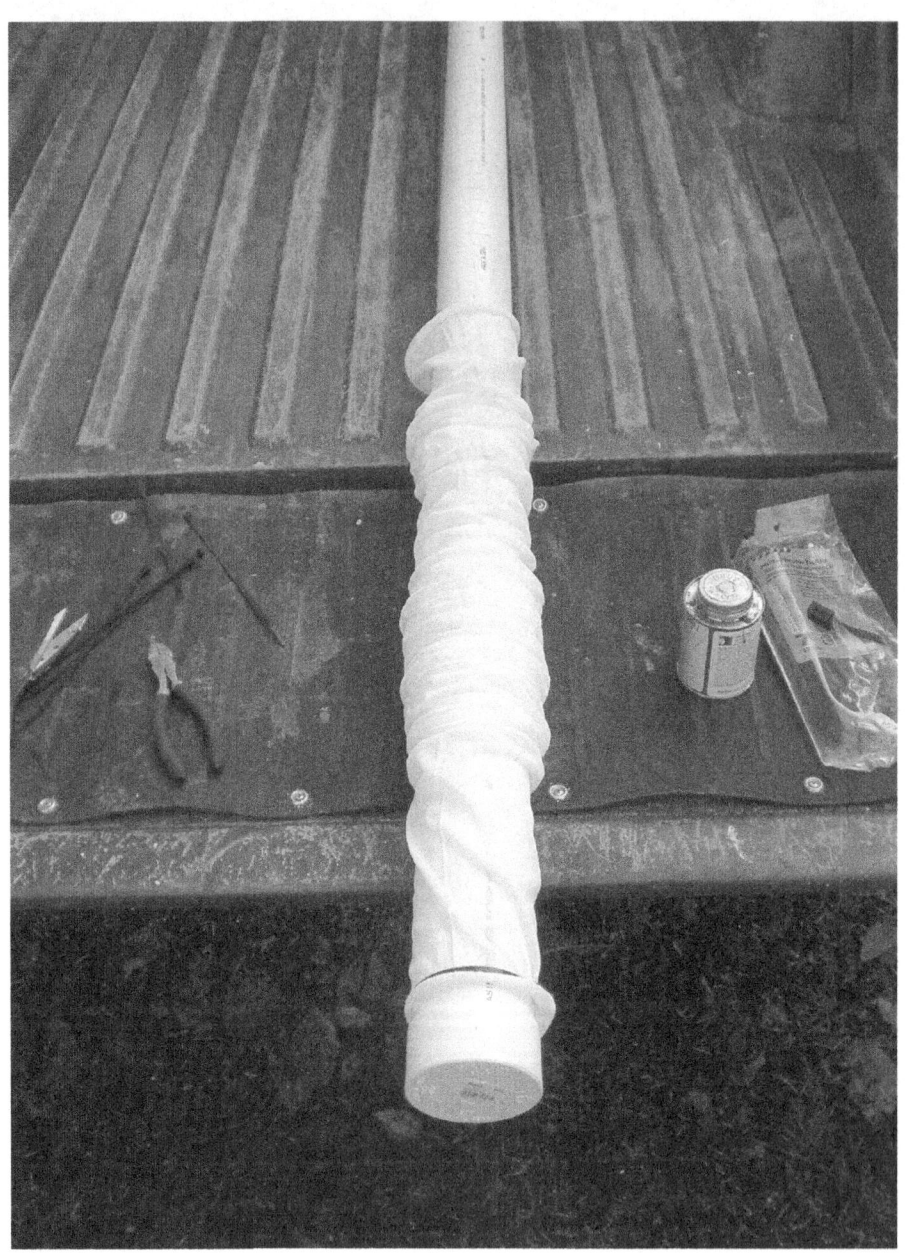

Once you have decided on a material to use for your sediment sock, you should install only enough to cover the slotted or drilled portion. If the sock is loose on a small diameter pipe, just wrap it snugly around the pipe and secure every 12" using plastic cable ties. Pay special attention to cable tie placement at each end as this is a possible point of sediment entry.

Bottom caped end of completed well screen. The screen in this photo is made from 3 inch thin wall PVC sewer and drain pipe covered with 4 inch diameter 'Drain Sleeve Filter Fabric Sock'. The sock fits slightly loose but this is not a problem, it will not effect filtering ability. Use plastic cable ties every 12 inches to secure sock.

Top bell housing end of completed well screen. This is the end that points up. The bell housing accepts the next segment of well casing pipe which is attached using PVC solvent cement. The 'Drain Sleeve Filter Fabric Sock' is secured at the throat of the bell housing with a plastic zip tie.

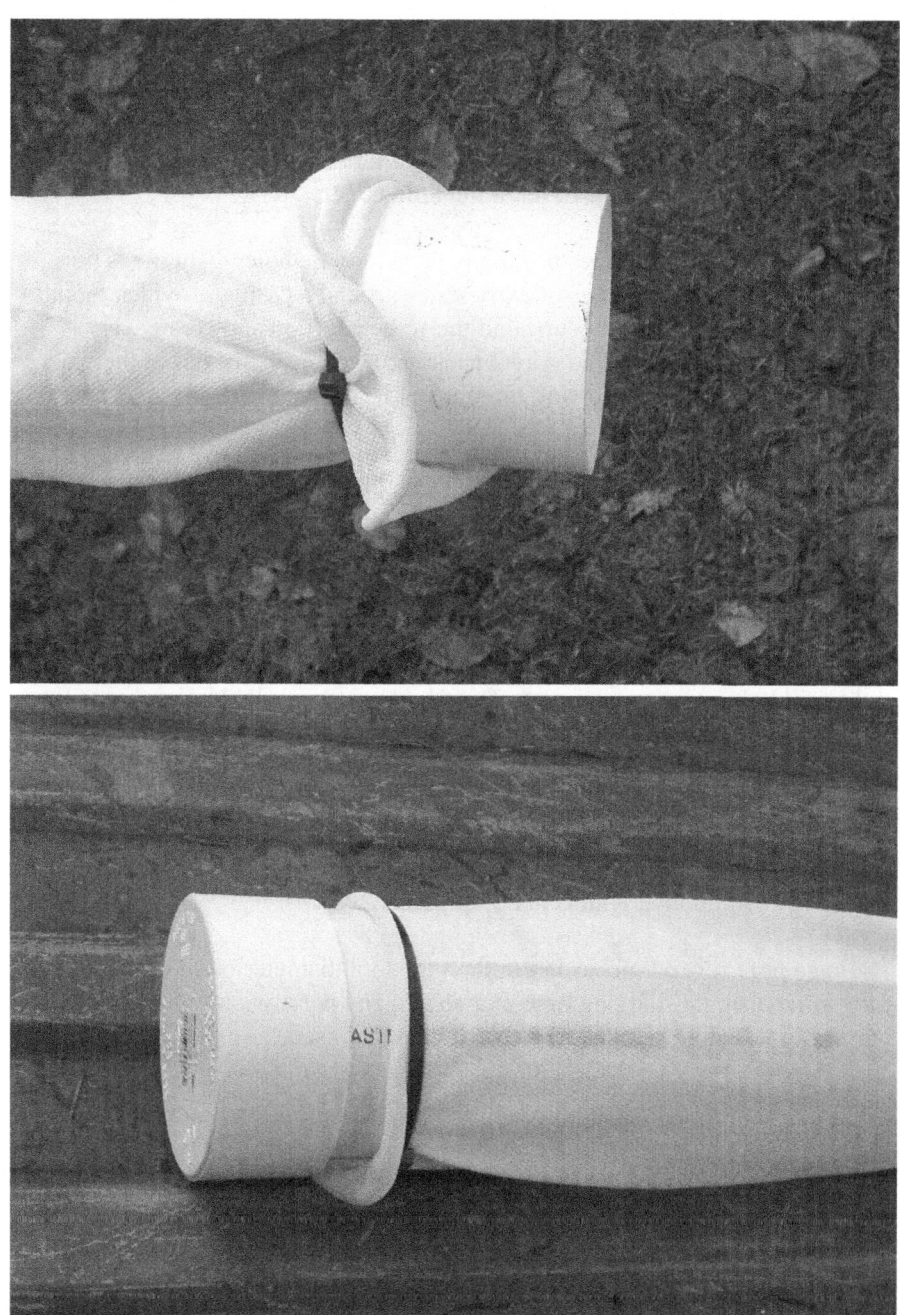

Top: bell housing at top end of well screen, filter sock and zip tie.

Bottom: capped bottom end of well screen, filter sock and zip tie.

Casing installation

Casing installation is a straight forward simple process assuming the bored hole is a few inches larger in diameter than the casing. If not you have miscalculated somewhere in the planning stages and need to; a) switch to a smaller diameter casing, or b) ream the hole larger.

In general, if the casing does not freely drop into the hole, that hole is not wide enough. Several inches of extra space around a casing provides room to pour gravel into the hole to surround the screen section and create what is known as a gravel pack. The gravel pack helps to keeps dirt out of the well while at the same time allowing more water to flow through the screen vents.

Concrete or grout is usually poured over the gravel to seal the well from surface run-off and other contaminants that might enter the water supply through any gap surrounding the casing. This layer of concrete or grout should extend all the way from the gravel pack to the surface possibly requiring many bags of concrete or grout mix if the well is deep.

You may have heard of the need for casings to be hammered or forced into tight well holes. Casings are often forced into rotary drilled holes due to the fact that rotary drilled holes are often bent (not perfectly straight down).

Most percussion and direct vertical hammering methods produce a straight bored hole. This makes installing the casing assembly easy as it is simply lowered into the hole, then the gap between the hole and casing is packed with gravel and sealed with grout or concrete.

Boring a hole several inches larger than the tool diameter often occurs naturally when using a sludging pipe, but augers and percussion tools tend to create a hole fairly close to the actual tool diameter used... so plan accordingly.

The first segment of well to be put down the borehole is the screen segment. It is lowered into the hole until 3-4 feet remain above ground. An assistant then holds the in-hole pipe while the next segment is attached and lowered. If no assistant is available the in-hole segment must be anchored to prevent it from falling into the hole.

PVC solvent cement is applied to the female bell housing joint of the screen tube, then the next casing segment is joined to it. This assembly is again lowered until 3-4 feet remain above ground. The process then repeats until the full depth of the borehole has been cased. Note: always install PVC screen and casing pipes with the bell housing facing up.

You should have a final exposed length of about 4-6 feet of casing protruding from ground level when complete. This can be trimmed to size later when pump installation work begins.

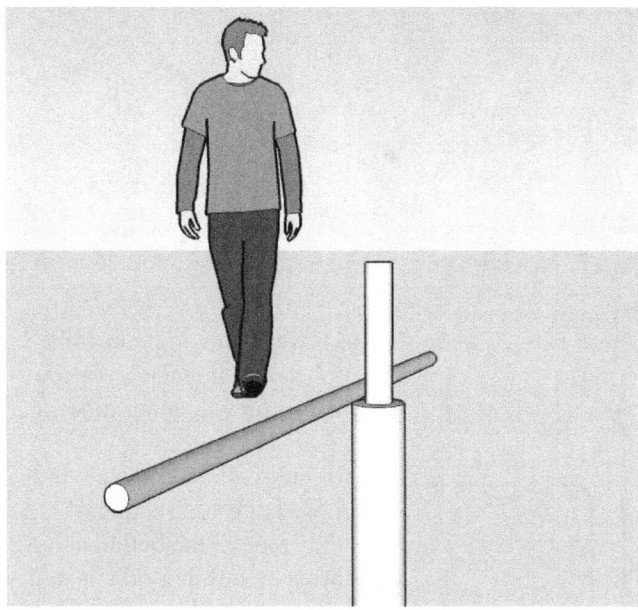

Note: in shallow wells with a high water table lightweight PVC casings will have a tendency to raise up out of the well due to mud around the casing being heavier than the water inside it. This produces pressure that will tend to lift up the casing.

While this phenomenon may seem an annoyance, it is actually welcomed in the sense that if you lose your grip on a segment of casing pipe and it drops into the hole... it will likely float back up and pop out of the hole a few feet.

After the casing is installed, further construction is required (such as grouting the bore hole perimeter etc). The casing must remain fixed for these tasks and the self-lifting effect will need to be stopped. To do this you simply tie the casing down once it hits the bottom of the bore hole. Have a few cinder blocks or a heavy bar next to the cased borehole. These will provide a sufficient anchor-to point. Just tie the casing to the weighted object.

For deep bore holes with a low water table, care must be taken during the casing install to not drop any unattached segments of pipe down the hole. You can cut an old bicycle tire inner tube in half to make an effective "rubber rope" for temporarily anchoring casings to the ground. It has a stretchy-sticky quality that holds with a simple wrapped knot. You can tie a piece of rope to the inner tube creating a longer safety line. The rubber end is easily wrapped and tied around casing pipe and will grip it in the event of a drop. Tie the inner tube around the throat of the upward facing bell housing flange to secure it during PVC solvent cement application. Attaching this safety line to each in-hole pipe segment being glued is low cost insurance against the job stopping headache of a dropped pipe.

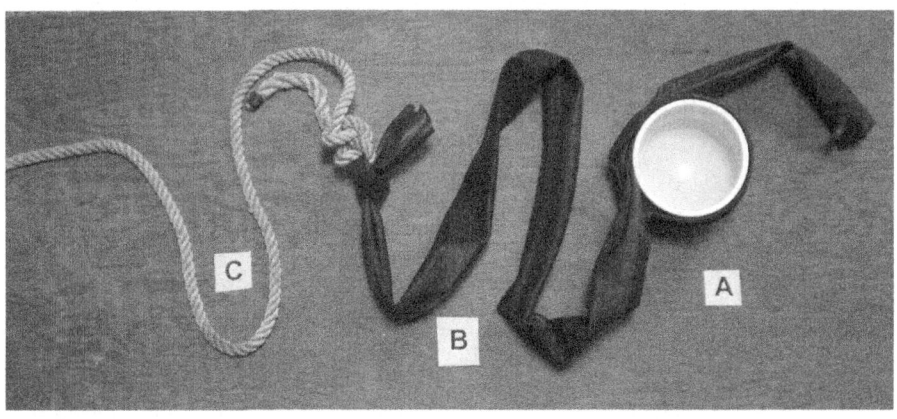

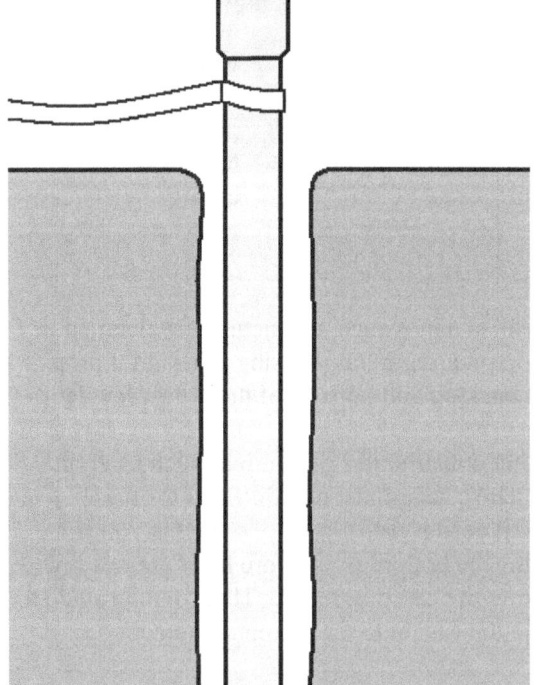

Above: casing pipe installation safety line device made from a bicycle inner tube.

Figure A: PVC cap shown to represent cross section of PVC pipe at the bell housing throat area. Inner tube is tied around the bell housing to support in-hole pipe during installation.

Figure B: black rubber bicycle tire inner tube cut in half.

Figure C: rope tied to inner tube and anchored to a stake located several feet away from the bore hole work area.

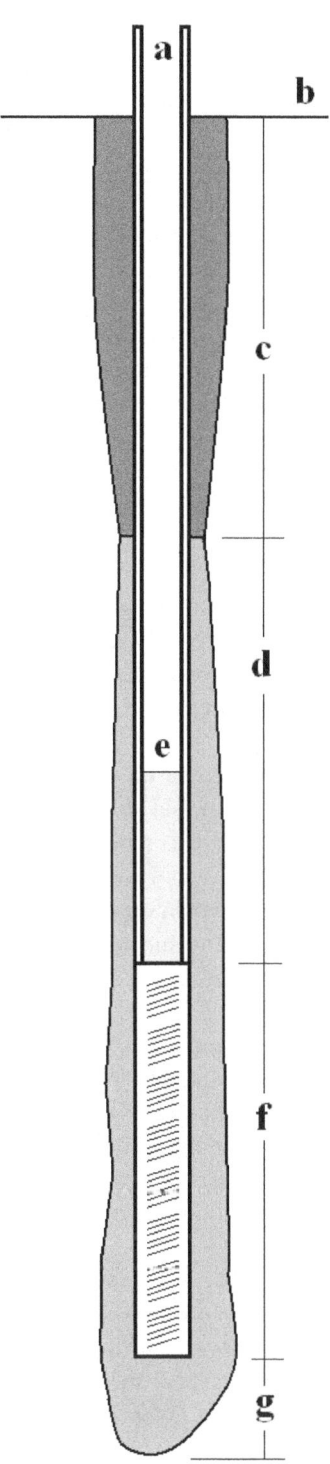

Figure a: complete PVC well casing usually has a few feet exposed at ground level for pump connection.

Figure b: two foot diameter concrete slab (4-6 inches thick) is poured around the exposed PVC casing at ground level. This seals the well from run off and surface contaminants.

Figure c: top 10 feet of casing should be sealed with concrete poured into bore hole around the casing. This is a sealing measure.

Figure d: this is the area between the sealed level and the gravel pack level. It will vary in length but must be filled with something, either concrete or gravel (larger diameter gravel is ok for this section).

Figure e: this is the "static head", or the waters normal resting "re-charged" level inside the casing pipe. It should be slightly higher than the screen segment for maximum flow.

Figure f: bottom segment of well. This is the screen submerged in ground water secreted by the natural water bearing formation that was drilled into. This section is encircled with small pea gravel.

Figure g: dump a bucket of gravel into the hole before screen installation begins. This will create a clean bed at the bottom of the bore hole for the casing screen to rest upon.

Static Head

The height of water column inside the well and available to the pump is usually less than the total well depth. This means that, when you reach a water bearing formation, water rarely shoots out of a well like you may have seen in the movies. The water in a borehole stays below ground level. So, the water level inside a 100 foot well casing may only reach up 20-40 feet. This means you need to lift (pump) your water an additional 50 plus feet to the surface.

The total volume of water that is available to the pump is called the "static head". The static head extends from the bottom of the well to the highest water level inside the well casing (at the waters normal resting "re-charged" state). This also demonstrates why well screen is only required at the bottom portion of the casing where the water is present.

Create Gravel Filter Pack

You will need several sacks of 1/8-1/4 inch diameter clean pea gravel. Exact quantity will depend on how large the gap between your casing and bore hole wall is. Avoid angular shaped gravel and anything over 1/4 inch as these will allow excessive sediment to gather near the screen.

The goal is to pour enough gravel into the bore hole to cover the screen section of tube by about 4 feet. Also keep in mind that the gravel level should not reach closer than 10 feet from the ground surface. This is a non-issue in deeper wells but, for example, say you have a shallow 25 foot well with a 10 foot screen section and you covered that 4 feet with gravel. You have then reached 14 foot with the gravel, leaving only 11 feet to the surface... just squeaking by within the aforementioned guidelines.

The reason for leaving the top 10 feet gravel-free is because that space will be sealed with concrete to keep out containments. Of course, there is room for flexibility within these standards. At the very least the gravel pack should extend 2 feet above the screen and the remaining space to the surface has concrete fill. Also, in deeper wells, you can fill any space remaining between gravel pack and grout layer with course gravel, concrete, or cement grout.

Use a weighted measuring tape to confirm when the top of the filter pack has reached sufficient level past the screen. You will have recorded earlier the length of screen, and each casing segment installed. These measurements, combined with the total depth of the bore hole, are essential for figuring out the correct gravel pack depth.

SAKRETE
ALL-PURPOSE
GRAVEL
Base material for posts and poles, walks, drives and patios to improve drainage
Net Wt. 60 lbs. (27.2kg)

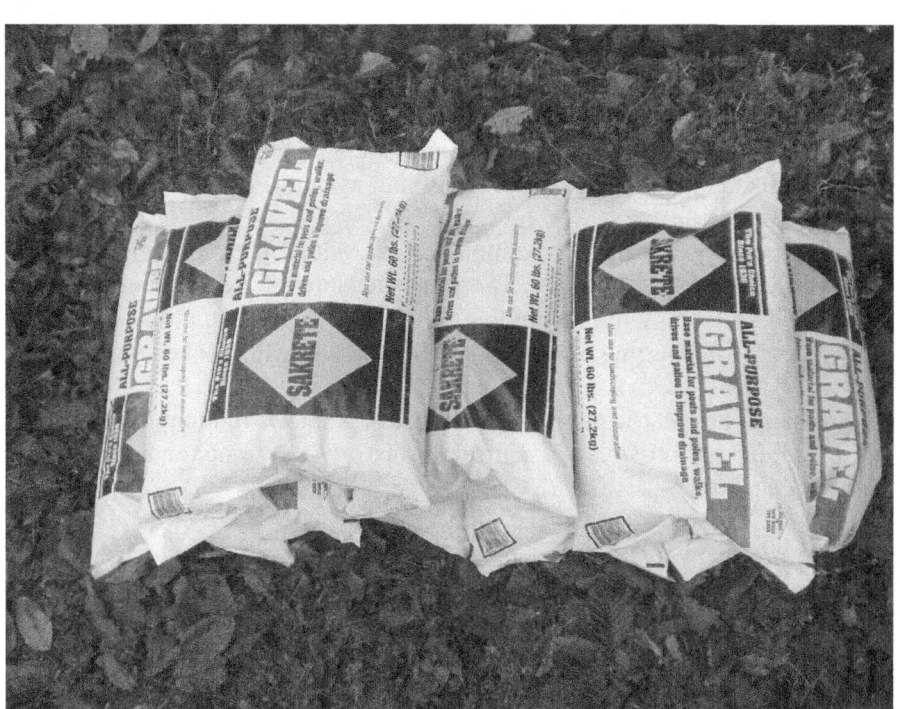

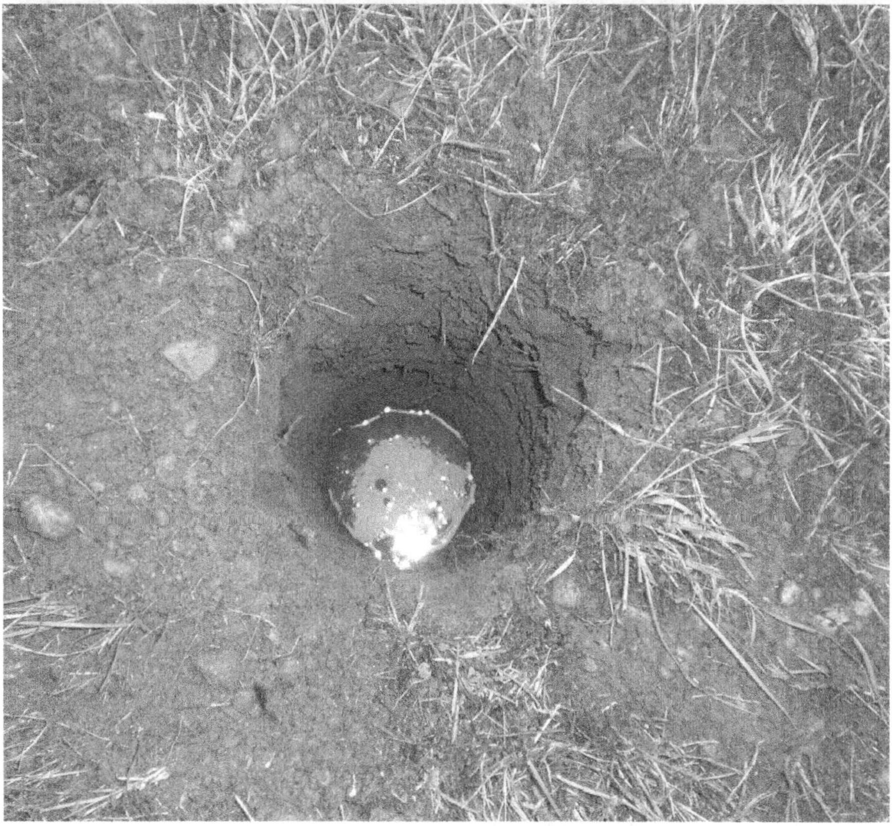

Pumps

The subject of pumps is wide ranging, but what can be considered low tech is a matter of debate. Hand operated pumps are about as low tech as you can get, but a hand operated pump places limitations on the users lifestyle.

Hand operated pumps need to be mounted over the well casing, this means they are usually located outside. This translates into a water hauling scenario and no indoor (supply line) plumbing. While it is possible to utilize a storage tank indoors, transferring gallons of water manually will still be required.

Upgrading to an electric pump and pressure tank makes running indoor supply line plumbing possible. Of course this requires electricity on site or some form of generator.

Solar or wind power can be harnessed in off-the-grid situations by using an electric pump and 12 volt deep cycle batteries connected through a power inverter. To reduce the pumps running time and maximize battery life, you really need a large pressure tank for this setup (not using much water also helps reduce pump run time). The large pressure tank will allow pressurized water to flow from your tap without the pump running. The more your pump runs in a solar/wind battery powered setup the less time your batteries have to recharge resulting in weak or no charge situations.

Using a non-pressurized water storage tank is also an option. Keeping a 30-100 gallon storage tank indoors in a temperature controlled environment will produce a readily available source of water to draw from without operating your well pump. Room temperature water can be heated or cooled as needed easier. This storage tank can be drawn from using a smaller electric pump or a hand pump. When the tank is empty, a gasoline generator can be used to power the main well pump and refill it.

Each type of pump has it's own installation requirements, and certain types of wells can only utilize certain types of pumps. Drive point wells for example, are too small in diameter to accept a submersible pump, and deep wells will require a pump able to lift water beyond 25 feet.

Left: shallow well jet pump. This type of electric pump can only be used on shallow wells less than 25 feet deep. The intake utilizes a check valve (on drive point wells), or a foot valve (on PVC cased wells).

The resting water level inside a well casing determines how deep the well is, not the casings in-ground length. Take a 200 foot deep well for example, with a high water level inside of just 10 feet below the grounds surface. This would be considered a shallow well and could be pumped with shallow well pumping methods. The opposite is also true. Take a 40 foot long in-ground well casing for example, with a 10 foot screen section at the bottom. Let's say the water level inside this casing is very low, rising only 5 feet up from the bottom of the well (or 35 feet from the surface). Since the cutoff depth to be considered a shallow well is 25 feet, this 35 foot deep well must be considered a deep well and pumped as such.

The nature of physics dictates that there are no exceptions to the way water is able to be lifted or pumped upwards. Even the act of mounting a pump 8 feet higher than ground level (inside a cabin, for example) will reduce that pumps depth capability by those 8 feet.

Above: shallow well jet pumps mounted on a pressure tank can be purchased as a single unit for space saving installation. Most electric well pump setups incorporate a pressure tank, either free standing, or with the pump mounted on it. The pressure tank allows a reserve of pressurized water to flow without the pump running each time a tap is turned on.

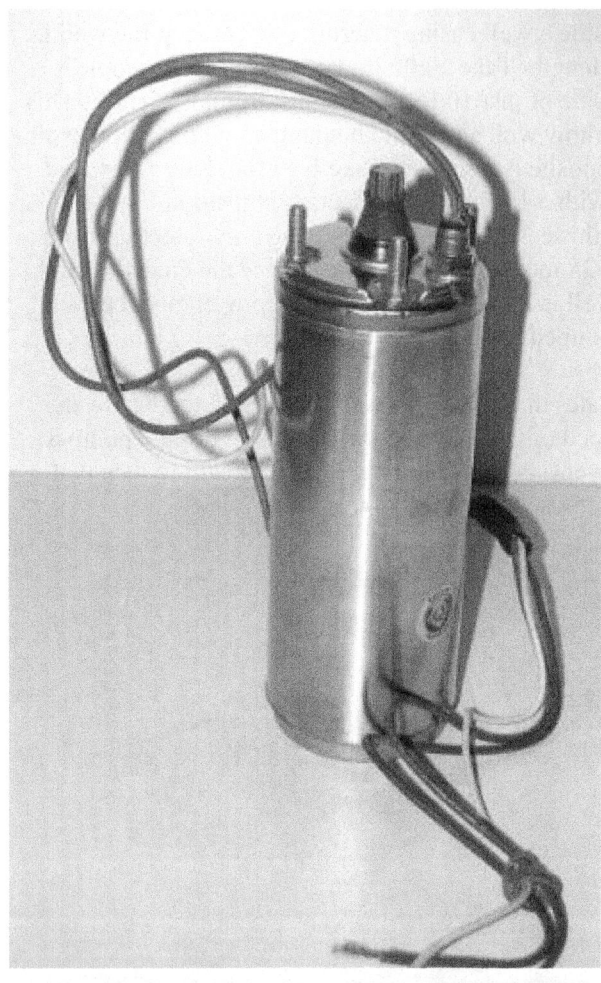

Submersible well pumps are lowered into a permanent position on a rope inside the well casing about 5-10 feet from the bottom of the well. At this level the pump is fully submerged and pumps water directly up a hose to the surface, exiting the casing wall at a point below the frost line through a 'pitless adapter'. This is the most common type of well pump in use today. While all other types of pumps pull water to the surface, the submersible pump pushes the water to the surface. Another benefit of location down the well tube is noise-free operation. This is appealing compared to surface jet pumps which are very loud.

Top: submersible well pump lowers into the casing on a rope. The wires and supply hose run up to the top of the casing and exit via PVC conduit and supply line pipe.

Bottom: shallow well jet pumps are commonly available with 1/4-1 hp electric motors.

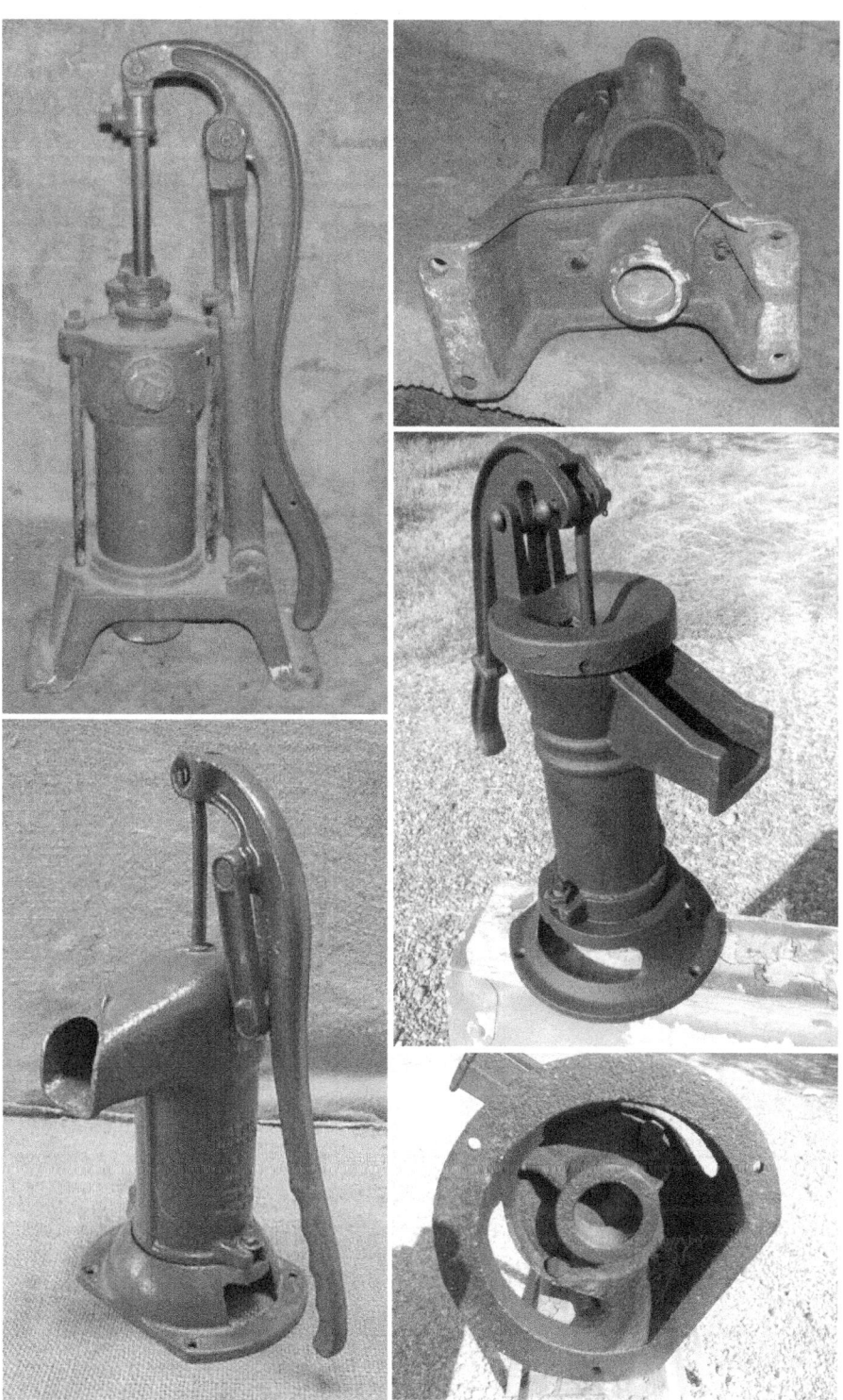

Hand pumps and pitcher style pumps have a threaded base for direct attachment to drive point wells. These type of shallow well pumps can lift water 15-20 feet to the surface.

Low Tech Hand Operated Well Pumps

Pump Design	Pumping Depth	Operation	Cost	Comments
Old fashioned type iron pitcher pump	20 feet	Hand operated manual pump	Low	1 GPM per 14 strokes 1-1/4 inch inlet
Old fashioned iron pedestal pump	23 feet	Hand operated manual pump	Low	1 GPM per 12 strokes 1-1/4 inch inlet
Oasis modern ABS pump	30 feet	Hand operated manual pump	Low	10 GPM per 50 strokes 2 inch inlet
Oasis modern ABS deep well pump	100 feet	Hand operated manual pump with in-casing drop cylinder	High	2 GPM per 50 strokes 2 inch inlet

Electric Jet Well Pumps

Pump Design	Pumping Depth	Operation	Cost	Comments
Shallow well jet pump	25 feet	Electric motor suction pump	Moderate	12.7 GPM, 30 PSI @ 15 feet
Deep well jet pump with packer ejector	20-90 feet	Electric motor suction pump	High	6.7 GPM, 30 PSI @ 40 feet
Deep well jet pump with two pipe system	200 feet	Electric motor submerged nozzle venturi	High	Rarely used, replaced by submersible

Electric Submersible Well Pumps

Pump Design	Pumping Depth	Operation	Cost	Comments
Submersible - pumps to fit 4 inch and larger casings	250+ feet - depth varies by model	Electric motor pushes water up from inside well casing	High	16 GPM @ 50 feet 6 GPM @ 200 feet

Water Flow Development of New Wells

Overview

Boring a hole in the ground and inserting a casing into it does not complete the well. The well must also be developed. This is the process of cleaning screen slots and getting the water flowing through and around the casing screen. Not properly or not fully developing a well often results in lower flow of water. The casing is inserted into an environment (the freshly bored well hole) that is undeveloped. Drilling mud, bentonite, fine gravel and other sediment is usually caked into the screen slots/holes and the surrounding water bearing rocks formations. You have to wash out the gunk that surrounds the casing and has impregnated the screen.

Well Development Methods

Bailing

Bailing out a well inside the casing is an effective method for cleaning out dirty water and fragments from new wells when other more advanced options (described below) are not available. This method fits within the title parameters of this book as it is about as primitive as you can get. Bailing will clean out the well but does little to promote well development. For this reason it should be followed with surging (see next page). All newly bored wells should be thoroughly bailed out removing large diameter debris prior to further well development.

Pumping

Over pumping is a method by which the well is pumped at a higher rate than expected for its future use. Since the new well will be full of dirty drilling fluid, a diaphragm pump must be used (a pump designed for solids that will not be damaged by sand and cuttings in the water). An average diaphragm pump will be powered by a small gasoline motor and offer the ability to pump out muddy water loaded with 1/2 inch rock fragments to a 25-30 foot depth. These pumps can usually be obtained from local equipment rental facilities for the day.

Surging

Mechanical surging is a method by which water in the well is forced in and out of the screen by raising and lowering a plunger apparatus within the well casing called a surge block. The surge block rig is most useful due to it's ability to be used as a supplement to the other well development techniques. Clean out muddy water and fragments first with a bailer before surging.

Surge block gasket disks can easily be fabricated from plywood disks and soft neoprene rubber, the same stuff wetsuits are made from.

Soft Neoprene is a more flexible blend of CR (Neoprene) and NBR (Nitrile) rubbers. It is often used for sealing and general gasket applications. Neoprene is the most commonly used industrial rubber sheet due to its all around excellent qualities and low cost. Neoprene rolls and sheets are available online in a variety of sizes.

Lengths of galvanized pipe and couplers should be used in series to provide the needed weight and reach for operating the surge block. Galvanized pipe comes in a standard 21 foot length pre-threaded. Use 1/2 inch pipe for a 2 inch casing surge block (each 21 foot segment weighs 18 lbs), and 3/4 inch pipe for a 4 inch casing surge block (each 21 foot segment weighs 24 lbs). Heavy pipe assemblies can be topped with a T-fitting and attached to a tripod and pulley to assist in operating.

Blowing

Blowing a well, while not low tech by any means, is a common practice used by modern well drilling contractors. This is a well development process that involves firing compressed air into the bottom of a new well to blow out dirty water and clear the screen.

It requires the use of the large trailer mounted compressors (70 to 100 cfm with about 80 to 100psi) available from your local tool equipment rental facility. Blowing the well lifts out water, clears the screen, and allows ground water to begin flowing.

You will need to run 1/2 inch PVC pipe down to about a few feet off bottom of the well. You must physically tie this pipe to the side of the casing (with rope) or the compressor will literally blow it out. Connect the compressor line (with appropriate fittings) to this pipe and turn the air on very slowly. You will need just enough to lift the water out the top of the well. If you put too much air down it, the air will pass by the water and not lift it. Extremely high air pressure can also tear the screen. Too little air and it will not lift the water. Once the water runs clear, turn the air off, wait about 2 minutes, and start the process all over until there is no more mud or silt. After the water is clear you can develop flow by turning the air on and off repeatedly (allowing water to surge out of the casing and screen).

Surge Block

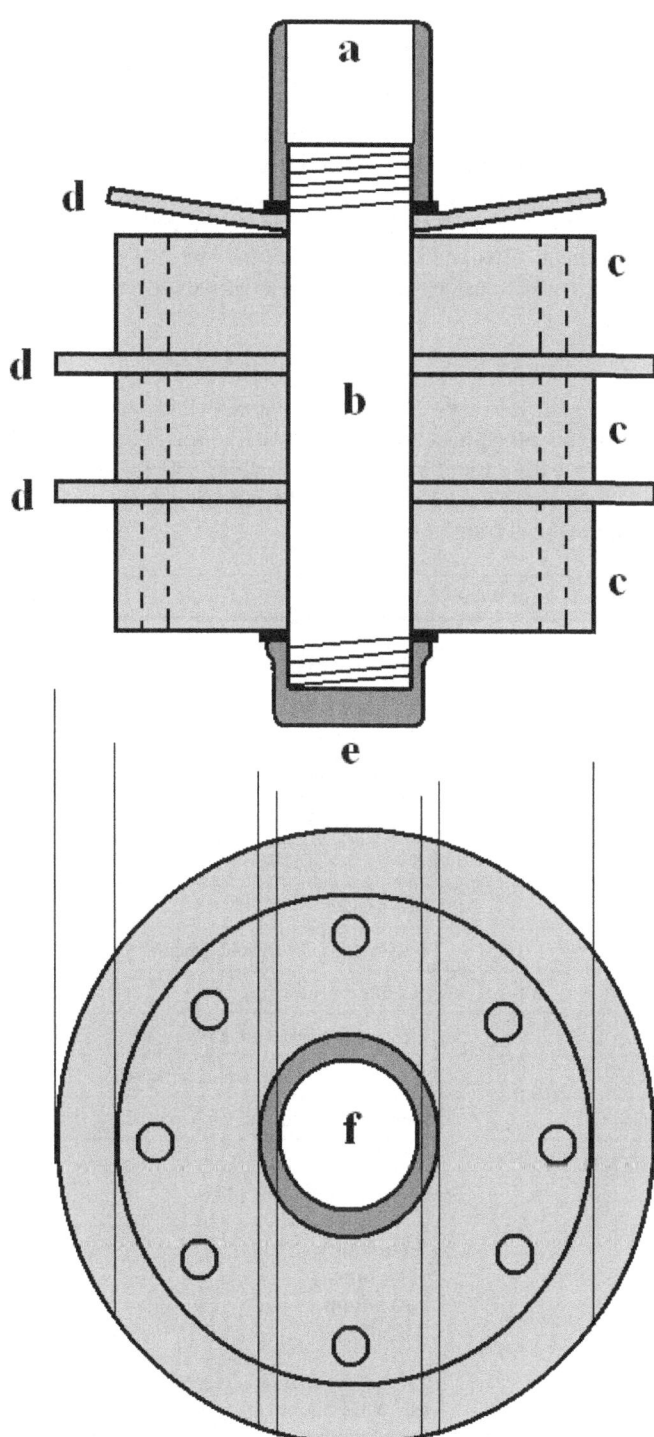

Figure a: 1/2-3/4 inch solid pipe coupler and washer. Assemble using thread locking fluid.

Figure b: 1/2-3/4 inch pipe nipple.

Figure c: 3/4 inch plywood discs cut 1/2 inch smaller than casing pipe inside diameter.

Figure d: soft neoprene disc cut to mach the casing inside diameter. Disc should make good contact but still be able to slide through.

Figure e: 1/2-3/4 inch pipe cap and washer. Assemble using thread locking fluid.

Figure f: top view of surge block shows vent holes through the plywood and neoprene discs. The vents will allow a small amount of water to pass preventing hydraulic lock.

Disinfecting Water Wells

New wells should be disinfected after the well development process. New wells often suffer from elevated levels of bacteria stirred up during the hole boring procedure. Wells should be tested and disinfected (if needed) every year thereafter. You need to test the water before and after disinfecting. Do-it-yourself test kits can be purchased for this. They usually include some fluids for adding to water samples that react in various shade of color. Some kits cost more and test for a wide spectrum of water conditions and contaminants, while some are specialized just testing PH etc. Local health departments offer test kits for a fee. Sometimes it's a simple water collection bottle with instructions on returning it to an office for testing.

Process

Determine how much water rests inside the well (casing diameter and static head). One quart household beach (per 50 gallons of water) is poured directly into the well. Add more water to the well via hose for 30 minutes. Leave chlorinated water inside the well for 12-24 hours. Remove the chlorinated water by pumping the well. Wait 5-10 days before re-testing the water for bacteria.

Problems Indicated by Odor or Taste of Water	Test For These
Sewer, gas, or swamp	Coliform, e-coli, nitrite, nitrogen, chloride, hardness, copper, iron
Musty	Bacteria, fluoride, chloride, hardness, copper, iron, manganese, uranium, arsenic
Sour, or metallic	Bacteria, fluoride, chloride, hardness, copper, iron, manganese, uranium, arsenic, color, turbidity, odor, sodium, detergents, cadmium, and zinc
Salty	Sodium and chloride
Fruity, or cucumber	Iron bacteria
Fishy, sweet, or perfume	Iron bacteria and volatile organics
Mothballs	Volatile organics
Rotten eggs	Hydrogen sulfide, or iron bacteria
Gasoline	Volatile petroleum (gasoline or MTBE)
Fuel oil, kerosene, or motor oil	Semi-volatile petroleum organics

New wells should be tested for:

- ° Bacteria
- ° Fluoride
- ° Chloride
- ° Hardness
- ° Copper
- ° Iron
- ° Manganese
- ° Uranium
- ° Arsenic

More advanced new well test includes all of the items to the left plus:

- ° Turbidity
- ° Sodium
- ° Uranium
- ° Lead

Checking the PH level is standard procedure for new wells and is required prior to disinfecting. The PH level must be between 6 to 7.5 for effective bleach disinfection. Always check water samples twice.

Hydrology Basics & Locating Ground Water

Overview

This chapter contains some geological descriptions of ground formations and other "not so low tech" instruction. Nevertheless it's good material for the would be well driller to read. To summarize the topic of locating ground water with a single all encompassing 'low tech' piece of advice: water depth will likely be reached at levels comparable to existing local wells. It's really as simple as that. Asking around to neighboring land owners to acquire this vital piece of knowledge before drilling a well is key to your success. Well depths are also recorded and kept on file by local and state agencies. You must have a target when drilling. The estimated water depth target gives you something to aim for and allows you to begin checking for water at that depth by dropping weighted lines, pumping, and so forth. Most ground water lies within the first 200 feet and the majority of wells are only 50-100 feet deep.

± **The Hydrologic Cycle**

Close to ninety seven percent of the earth's fresh water (not including the fresh water frozen in the polar ice caps and glaciers) is located underground. Most of the water that ends up in a well originates from precipitation on the earth's surface. The act of it passing through various ground formations on it's way to your well naturally filters the water.

The constant movement of water above, on, and below the earth's surface is the hydrologic cycle. The concept of the hydrologic cycle is that precipitation returns again to the atmosphere by evaporation and transpiration. Some basic knowledge of hydrology is an important tool most well drillers should possess.

Three-fourths of the earth's surface is covered by ocean water. Direct radiation from the sun causes water at the surface of the oceans to change from a liquid to a vapor (evaporation). Water vapor rises in the atmosphere and can accumulate as clouds. When the clouds accumulate enough moisture and conditions are right, the water is released in the form of precipitation (rain or snow). The rain water that is not evaporated immediately or carried off by streams sinks into the ground. Practically all ground water encountered in well drilling is derived from precipitation.

In the eastern United States where rainfall is plentiful, the yearly average varies from 30 to 50 inches. More than 70 inches a year falls on the Mississippi Delta below New Orleans, and along the Gulf coast from near Mobile, Alabama, to Tallahassee, Florida. A nearly equal amount falls in the higher mountains of western North Carolina and eastern Tennessee, along the coast of North Carolina, and in the Adirondack and White mountains. In the Gulf and South Atlantic States the rainfall is between 50 and 60 inches a year; in the New England, Central Atlantic, and Ohio River States, between 40 and 50 inches; in the upper Mississippi and Great Lake States, between 30 and 40 inches; and in northwestern Iowa and most of Minnesota, between 20 and 30 inches.

In the western part of the United States the distribution of the rainfall is much more irregular. Westward from a line drawn through the eastern part of the Dakotas, middle Nebraska, western Kansas, and central Texas the rainfall quickly decreases to less than 20 inches yearly, all of the Great Plains regions are characterized by small rainfall. In the Black Hills, the Bighorn Mountains, and the higher sections of the main chains of the Rocky Mountains the rainfall is 20 or 30 inches yearly; but in the high Sierra, the Cascades, and the Coast Ranges it is 70 inches or more. In the Great Basin region, between the Sierra Nevada and the Wasatch Mountains, the rainfall is less than in any other section of the country, in places being as low as 2 or 3 inches per year. The most rain received in the United States falls upon Mt. Waialeale, Kauai, Hawaii with an average of 460 inches per year of rainfall. The most rain received in the lower forty eight contiguous United States is Washington States Olympic Peninsula Rainforest receiving an average 170 inches of rain per year.

° Groundwater supplies are chiefly fortified by precipitation, meaning the average low tech bored well will achieve a much higher rate of success in a rain prone zone.

° In most places it is necessary to penetrate some distance below the earth's surface in order to reach a zone saturated with water, the actual depth depends on the amount of precipitation, the specific geology, and topography of that particular well drilling location.

° This drilling depth required is shallow in regions of high rainfall (and deep in dry regions); and in general it is shallow in valley bottoms and deeper on high ground. In some locations, as at springs and in marshy lands, the plane of saturation coincides with the surface, but the existence of ground water at the surface is due to exceptional conditions. High water tables (less than 25 feet) are not recommended for use as drinking water sources due to possible contamination.

In order that a well may flow, traditionally, it is believed that the following conditions must be satisfied:

1. There should be sufficient rainfall.

2. There should be relatively porous beds suitably exposed to collect and transmit the water.

3. There should be less porous or relatively impervious layers that confine the water collected.

4. The level of the ground water at the source should be at a sufficient height above the mouth of the well to compensate for the loss of head due to resistance and leakage.

The arrangement of the factors which produce a flow are by no means constant. These factors vary considerably, and relatively new combinations occur constantly.

± Water Bearing Formations

Geologic knowledge has perhaps no greater practical value than in the study of the occurrence of underground water and in the selection of sites for drilled wells. The locating of wells is subject to so much superstition and guesswork among drillers and prospective well owners that it seems worth while to point out a few general principles that should be observed.

Many of the details that influence the presence of underground water can be intelligently determined only by means of an examination of the region in which the well is to be sunk and a careful analysis of observations of other wells put down in the locality. The observing well driller should also take into account the surface drainage of the region, the presence or absence of forest lands which may influence the rate and amount of run-off, the existence of springs as evidence of surplus water at given points, and the depth and character of soil overlying the rock. Consider the structure of the soil cover foliage.

Ground water has an intimate relation with the surface drainage and the topography. Even after the most thorough investigation of the surface geology, however, there is always a certain degree of chance in the results of the first well drilled in a region.

The variations from conditions predicted are, as a rule, small, or even insignificant, but occasionally they are great. Therefore the conclusions of a competent geologist are to be considered by the well driller as a moderately safe guide in a new region, one which will enable him to determine roughly the possibility of obtaining water, the method of drilling to be employed, and the length and kind of casing that will be required. More reliable geologic inferences, however, can be drawn only after some drilling has been done in the region; so that the geologist and the driller must supplement each other.

± Sands and Gravels

Beds of sand and gravel are very porous, as much as 30 per cent of the volume of some of them being made up of free space, so that saturated layers that are penetrated by wells yield copious supplies of water. This water is in most places of good quality, but in some wells it is greatly mineralized with salts derived from the more soluble rock fragments that constitute the deposits.

In passing downward through sands, especially the finer varieties, surface water may be naturally filtered so that substances with which it may have been polluted are removed. In coarse sands and in gravels the water passes downward more rapidly and the conditions are less favorable for its filtration, so that it may remain polluted. In general, however, water derived from sands and gravels that lie a considerable distance below the surface is pure.

Because of the readiness with which sands and gravels yield their water, wells that are sunk close together in them may affect one another, the deeper wells, or those which derive their supply from the sands at the lowest points, drawing the water from the shallower wells. The readiness with which the water moves is also the cause of important fluctuations in the level of the surface of the saturated zone, as the water level may fall rapidly after wet seasons. To procure permanent supplies a well should penetrate below the level to which the water surface falls in the driest season.

± Clay

Pure clay is nearly impervious to water and contains little or none that can be utilized as a source of supply. Water is frequently reported in clay, but as a rule it comes from more or less sandy layers within a clay bed. In some places sand that approaches clay in fineness and that is sometimes mistaken for clay yields considerable amounts of water. Clay is of great importance in connection with underground water, not directly as a water-bearer, but as a confining layer to porous sands from which it prevents the water from escaping.

Because of the fineness of clay the water it contains comes into contact with a relatively large amount of mineral matter and may become mineralized, especially by lime and salt.

When it is necessary to obtain water from clay the well should be as large as practicable and should be carried deep enough to provide ample storage capacity, for clay yields only a small amount of water, and that very slowly. Dug wells are usually most satisfactory where clay is near the surface, but should be carefully covered and protected from pollution.

± Till

Till is a heterogeneous mixture of clay, sand, gravel, and boulders. In texture it ranges from very pervious to impervious, according to its content of sand or of clay.

In few places is it definitely bedded. Water generally occurs in it in minute, more or less tubular, channels, but occasionally is distributed through interstratified sandy beds. In the finer, more loamy phases the supply is not abundant, but the coarser portions of the material furnish water more plentifully.

In the aggregate, till yields a large amount of water, and where it is sufficiently thick it forms a most convenient and accessible source of supply, but because its material is so irregularly disposed the success of the wells varies greatly. Water is usually found in it close to the surface, but better water can be obtained by casing off these upper water-bearing beds and extending the wells to greater depth.

In general, wells of large diameter, similar to those dug in clay, are most satisfactory, for the surface of the till exposed inside the well—that is, the surface from which water can enter the well—is much greater than that exposed in a bored well, and the larger wells also have greater storage capacities. Dug wells may therefore utilize small supplies which would be insufficient for bored wells.

The water of till is generally uncontaminated, because in its downward passage through the clay, of which till is in part composed, it is subjected to thorough filtration. If the water does become polluted, however, it may retain its dangerous character for a long time and for considerable distances.

Till is widely distributed over the northeastern and north-central parts of the United States.

± Sandstone, Conglomerate, and Quartz

Sandstone is, on the whole, the best water bearer of the solid rocks. Under the most favorable conditions it is saturated throughout its extent below the ground-water level and yields water freely wherever it is struck by the drill, although some of the fine-grained sandstones yield it less readily.

Water derived from sandstone is of better average quality than that obtained from any other material except sand and gravel. It is seldom polluted, and the wells supplying it can generally be safely used if they are cased to keep out surface water. Wells obtaining water from sandstone are usually drilled, although where the sandstone is very near the surface dug wells are common.

± Shale

Shale, like clay, is a poor water bearer, but may yield water from bedding, joint, and cleavage planes, and other crevices. It is most important, however, as a confining layer to prevent the escape of water from porous sandstone that may be interbedded with it. The water in shale is reached by deep wells and is generally uncontaminated, though it may be salty or bitter or otherwise undesirable.

± Limestone

The water of limestone occurs mainly in open channels and caverns which have been formed by the solvent action of water passing along joint or bedding planes. These caverns and water channels are very irregularly distributed and their location can seldom be determined by examining the surface, but most deep wells in limestone regions encounter one or more such passages at relatively slight depth. Adjacent wells sunk in limestone, even if only a few feet apart, may obtain very different results, as a difference of a foot or two in position may mean missing a water channel.

The water of limestone regions contains dissolved calcium carbonate (commonly called carbonate of lime) which makes it hard, but usually it is not otherwise heavily mineralized. It is, however, likely to be polluted due to the fact that much of the water of the underground streams in limestone has found its way downward through sink holes and carries with it more or less surface wash.

± Lake and Stream Deposits

Along the northern portion of the Atlantic coastal plain there are extensive lake and stream deposits of clay, sand, and gravel, which are largely drawn upon for domestic water supply. This water is usually plentiful and good.

In the lower Mississippi Valley the fine silts of the flood-plain deposits are saturated with water a short distance below the surface and furnish abundant supplies to small wells. The chief drawback to shallow wells in this material is the unusually small mesh of the screens that are necessary to exclude the fine sand from the wells. In most places, however, a lens of coarser sand or gravel can be located, and this can be drawn on for water supply without danger of clogging the screen.

In many parts of the arid region of the western United States there are great basins or valleys which are deeply filled with sediments brought down by streams and deposited in former lakes. In some places these unconsolidated sands and silts contain water at shallow depths, and they form the chief source of supply throughout much of the Great Basin region of Utah, Nevada, southeastern Oregon, and southeastern California. In the more favorably situated of these desert valleys, notably in the Coachella Valley, in southeastern California, the deeper water of the unconsolidated deposits may be under sufficient artesian head to yield flowing wells.

In the great central valley of California and in the coastal slopes of the Pacific States there are also deep alluvial deposits that are of great importance as sources of underground water. Those of the coastal plain of southern California have been extensively tapped to obtain water for irrigation, and water is also drawn from similar deposits around Puget Sound.

Many smaller deposits of alluvial material at the bases of the western mountain ranges also yield valuable supplies of water to shallow wells.

± Well Drilling Methods (Isaiah Bowman), Water Supply Paper 257, United States Department of the Interior, Geological Survey, 1911

Made in the USA
Las Vegas, NV
25 April 2025

21309095R00075